LAB MANUAL
TO ACCOMPANY

ELECTRONIC
COMMUNICATIONS
SYSTEMS
FUNDAMENTALS
THROUGH
ADVANCED

Wayne Tomasi
Mesa Community College

Cheryl Tomasi

PRENTICE HALL, Englewood Cliffs, New Jersey 07632

Editorial/production supervision and
interior design: **Lisa Schulz Garboski**
Cover design: **Diane Saxe**
Manufacturing buyer: **Peter Havens**

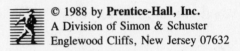
Printed in the United States of America

10 9 8 7 6 5 4 3 2 1

ISBN   0-13-250853-2

PRENTICE-HALL INTERNATIONAL (UK) LIMITED, *London*
PRENTICE-HALL OF AUSTRALIA PTY. LIMITED, *Sydney*
PRENTICE-HALL CANADA INC., *Toronto*
PRENTICE-HALL HISPANOAMERICANA, S.A., *Mexico*
PRENTICE-HALL OF INDIA PRIVATE LIMITED, *New Delhi*
PRENTICE-HALL OF JAPAN, INC., *Tokyo*
SIMON & SCHUSTER ASIA PTE. LTD., *Singapore*
EDITORA PRENTICE-HALL DO BRASIL, LTDA., *Rio de Janeiro*

CONTENTS

EXPERIMENT 13 —CLASS C, AM DSBFC MODULATOR   95

EXPERIMENT 14 —LINEAR INTEGRATED CIRCUIT AM MODULATORS   103

EXPERIMENT 15 —TUNED AMPLIFIERS   111

EXPERIMENT 16 —AM PEAK DETECTOR   117

EXPERIMENT 17 —AUTOMATIC GAIN CONTROL (AGC)   125

EXPERIMENT 18 —LINEAR INTERGRATED CIRCUIT PHASE LOCKED LOOP   131

EXPERIMENT 19 —PHASE-LOCKED LOOP FREQUENCY SYNTHESIZER   141

**EXPERIMENT 20 —LINEAR INTEGRATED CIRCUIT BALANCED
                          MODULATOR   149**

Contents

EXPERIMENT 37 —PCM/TDM SYSTEM DESIGN USING COMBO CHIPS 285

EXPERIMENT 38 —PCM/TDM LINE ENCODING 293

EXPERIMENT 39 —BINARY SUBSTITUTION ENCODERS 301

EXPERIMENT 40 —BIPOLAR RETURN-TO-ZERO ALTERNATE MARK INVERSION ENCODER/DECODER DESIGN 311

EQUIPMENT LIST

1. Protoboard
2. dual dc power supply ($+/-$ 20 Vdc)
3. Medium frequency signal generator (1.5 MHz)
4. Function generator (100 kHz)
5. Audio signal generator
6. Oscilloscope (10 MHz)
7. Voltmeter (dc and ac)
8. Resistor substitution box
9. Capacitor substitution box

PARTS LIST

Resistors	Capacitors	Inductors
1 - 22 ohm	1 - 30 pF	1 - 3 mH
3 - 51ohm	5 - 0.001 μF	1 - 7.1 to 12.5
10 - 100 ohm	1 - 0.0022 μF	μH
1 - 470 ohm	1 - 0.0047 μF	
11 - 1k ohm	1 - 0.001 μF	**Transistors/diodes**
2 - 1.2k ohm	5 - 0.002 μF	
1 - 2.2k ohm	2 - 0.01 μF	1 - 2N3904
1 - 2.7k ohm	1 - 0.022 μF	1 - 2N5457
2 - 3.9k ohm	1 - 0.05 μF	1 - 1N914
5 - 4.7k ohm	3 - 0.1 μF	1 - 1N4004
1 - 6.8k ohm	2 - 0.5 μF	
1 - 8.2k ohm	1 - 1 μF	**Integrated Circuits**
4 - 10k ohm	1 - 10 μF	
1 - 18k ohm		1 - MF10
1 - 22k ohm		1 - LS123
1 - 33k ohm		1 - LS398
2 - 47k ohm		1 - LM741C
2 - 100k ohm		1 - XR-2206
1 - 1M ohm		1 - XR-2211
1 - 1k ohm variable		1 - XR-2212
1 - 5k ohm variable		1 - 2913/14
2 - 10k ohm variable		2 - CD4027B
		1 - CD4069B
		1 - CD4071B
		2 - CD4081B
		1 - IM6402

PREFACE

This laboratory manual contains 40 experiments that are intended to be used with Electronic Communications Systems: Fundamentals Through Advanced, although the laboratory experiments are appropriate for almost any comprehensive electronic communications text. This manual includes experiments on fundamental communications topics such as filters, signal analysis, harmonic and intermodulation distortion, and oscillators; various modulation techniques such as AM, FM, and sideband transmissions; communications receivers and receiver circuits; and phase locked loops. There are also several experiments addressing data communications hardware, such as parity generation/checking and UARTS; and digital communications techniques, such as FSK and BPSK. Nineteen experiments incorporate modern special function linear integrated circuits. Also, several design experiments have been included into the manual to challenge the student's ability to design and test original circuits to a given set of specifications. The intent of this laboratory manual is to enhance the material that is covered in class lecture and to reinforce the student's understanding of basic electronic communications concepts. Experiments 4 through 40 cover concepts in the same order that the material is presented in the text. Experiments 1, 2, 3, and 34 deal with the concepts of filter circuits (both active and passive) which are commonly used in electronic communications systems but often covered in prerequisite classes on signal or network analysis. Experiments 1, 2, 3, and 34 can be done in the laboratory during the beginning part of the semester, thus giving the instructor sufficient class time to cover preliminary topics from Chapter 1.

Each experiment is divided into sections and each section builds from the previous sections. However, it is not necessary that all sections from each experiment be done. A certain degree of redundancy has been purposely incorporated into the experiments. Several concepts are presented in more than one experiment. For example, various aspects of filter concepts and tuned amplifiers are covered in several different experiments. If experiments 1, 2, 3, and 15 were done, sections from other experiments dealing with filter concepts and tuned circuits could be omitted.

The type of circuits, component values, and the number and type of special purpose integrated circuit chips used in these experiments were carefully selected to require the minimum number of parts and still retain the fundamental concepts the experiments are intended to present.

Wayne Tomasi
Cheryl Tomasi

PASSIVE RC FILTERS

REFERENCE TEXT: James Harter and Paul Lin, *Essentials of Electric Circuits*, 2nd edition (Englewood Cliffs, N.J.: Prentice-Hall, 1986).

Chapter 29, Introduction to Filters.

OBJECTIVES

1. To observe the frequency response characteristics of passive RC filters.
2. To observe the frequency response characteristics of lowpass, highpass, bandpass, and bandstop filters.

INTRODUCTION

In electronic communications circuits, it is often necessary to separate a single frequency or a specific band of frequencies from a composite waveform. Frequency separation is accomplished with filter circuits. In essence, a filter is a circuit that either amplifies or attenuates a particular range or band of frequencies more than others. There are four basic types of filters: lowpass, highpass, bandpass, and bandstop. The filter name describes its frequency response: a lowpass filter passes relatively low frequencies, a highpass filter passes relatively high frequencies, a bandpass filter passes a band of frequencies, and a bandstop filter stops or blocks a band of frequencies. In addition, filters can be classified as either active (i.e., filters with voltage gain) or passive (i.e., filters with no voltage gain). A passive filter generally introduces a signal loss for all frequencies, including those that fall within its passband. The loss incurred by a frequency that falls within the passband of the filter is called insertion loss. The simplest filters are constructed from resistors and passive frequency dependent components such as capacitors and inductors. In this exercise, the four basic types of filters are constructed using resistors and capacitors, and their frequency response curves are examined. The number of stages in a filter determines how rapidly its output voltage decreases (rolls off) with frequency. A stage is often referred to as a pole. The filters examined in this experiment are single-pole filters. The block diagram for the test circuit used in this experiment is shown in Figure 1-1.

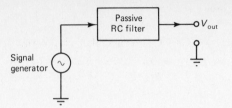

FIGURE 1-1 Passive RC filter test circuit

MATERIALS REQUIRED

Equipment:

1 — protoboard
1 — medium-frequency signal generator (1 MHz)
1 — standard oscilloscope (10 MHz)
1 — assortment of test leads and hookup wire

Parts List:

2 — 10 k-ohm resistors
1 — 47 k-ohm resistor
2 — 100 k-ohm resistors
2 — 0.001 μF capacitors
1 — 0.002 μF capacitor
1 — 0.0047 μF capacitor

SECTION A Lowpass Filter

In this section the frequency response of a passive RC lowpass filter is examined. The schematic diagram for the lowpass filter circuit used in this section is shown in Figure 1-2. The output signal is taken across capacitor C_1. At relatively high frequencies, the reactance of C_1 is extremely low, and at relatively low frequencies, the reactance of C_1 is extremely high. Therefore, C_1 looks like a short circuit to high frequencies and an open circuit to low frequencies. Thus, high frequencies are shorted to ground, and low frequencies appear across C_1 (hence, a lowpass filter). The break frequency for the filter is the frequency at which the output voltage decreases by a factor of 0.707. The term break frequency implies that the output amplitude breaks into a different response at that frequency. For a lowpass filter all frequencies below the break frequency produce the same output voltage for a given input voltage, and the output voltage for all frequencies above the break frequency decreases at a rate equal to 20 dB per decade (i.e., 6 dB per octave). The break frequency for a lowpass filter is often called the upper cutoff frequency. A factor of 0.707 was chosen as the reference value because a voltage decrease of 0.707 causes a reduction in power of 50% (i.e., $P = E^2/R$ and $0.707^2 = 0.5$).

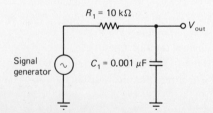

FIGURE 1-2 Passive RC lowpass filter

Procedure

1. Construct the RC lowpass filter circuit shown in Figure 1-2.
2. Calculate the upper cutoff frequency for the filter using the following formula. (When $XC_1 = R_1$, the signal voltage is distributed equally across R_1 and C_1.)

$$F_U = \frac{1}{2\pi RC}$$

   where F_U = upper cutoff frequency
   $R = R_1$
   $C = C_1$

3. Measure the actual upper cutoff frequency by setting the signal generator output voltage to 4 Vp–p and varying its output frequency from 100 Hz to 100 kHz.
4. Calculate the percentage error between the calculated and actual upper cutoff frequencies using the following formula.

$$\% \text{ error} = \frac{\text{measured value} - \text{calculated value}}{\text{calculated value}} \times 100$$

5. Plot the frequency response curve for the RC lowpass filter on semilog graph paper.
6. Measure the insertion loss of the filter using the following formula.

$$IL = \frac{V_{\text{out}}}{V_{\text{in}}}$$

   where IL = insertion loss
   V_{out} = filter output voltage
   V_{in} = filter input voltage

7. Design a lowpass filter with an upper cutoff frequency $F_U = 1600$ Hz. Use a resistance $R = 10$k ohms.
8. Construct the lowpass filter circuit designed in step 7, and plot its frequency response curve on semilog graph paper.

SECTION B   Highpass Filter

In this section, the frequency response of a passive RC highpass filter is examined. The schematic diagram for the highpass filter circuit used in this section is shown in Figure 1-3. The output signal is taken across R_1. At relatively high frequencies, the reactance of C_1 is extremely low, and at relatively low frequencies, the reactance of C_1 is extremely high. Therefore, C_1 looks like a short circuit to high frequencies, and all the signal generator's output voltage is dropped across R_1. C_1 looks like an open circuit to low frequencies and thus prevents them from reaching the output. For a highpass filter, all frequencies above the break frequency produce the same output voltage for a given input voltage, and all frequencies below the break frequency decrease at a rate equal to 20 dB per decade. The break frequency for a highpass filter is often called the lower cutoff frequency.

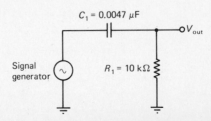

FIGURE 1-3 Passive RC highpass filter

Procedure

1. Construct the RC highpass filter circuit shown in Figure 1-3.
2. Calculate the lower cutoff frequency for the filter using the following formula.

$$F_L = \frac{1}{2\pi RC}$$

   where F_L = lower cutoff frequency
   $R = R_1$
   $C = C_1$

3. Measure the actual lower cutoff frequency by setting the signal generator output voltage to 4 V p–p and varying its output frequency from 100 Hz to 100 kHz.
4. Calculate the percentage error between the calculated and actual lower cutoff frequencies.
5. Plot the frequency response curve for the highpass filter on semilog graph paper.
6. Measure the insertion loss of the filter.
7. Design a highpass filter with a lower cutoff frequency F_L = 34 kHz. Use a resistance R = 1k ohm.
8. Construct the highpass filter circuit designed in step 7, and plot its frequency response curve on semilog graph paper.

SECTION C   Bandpass Filter

In this section, the frequency response characteristics for a passive RC bandpass filter are examined. In essence, a bandpass filter is the combination of a lowpass and a highpass filter. The schematic diagram for the bandpass filter circuit used in this section is shown in Figure 1-4a. The output signal is taken across the parallel network of R_2 and C_2. At relatively low frequencies, C_2 looks like an open circuit, and at relatively high frequencies, C_1 looks like a short circuit. The low-frequency and high-frequency equivalent circuits are shown in Figures 1-4b and 1-4c, respectively. The lower break frequency for the bandpass filter is the break frequency for the highpass filter, and the upper break frequency for the bandpass filter is the break frequency for the lowpass filter. The filter shown in Figure 1-4a has two stages (and thus two poles); however, the poles are at different frequencies. Consequently, each stage acts like a separate filter and has little effect on the frequency response of the other stage.

Procedure

1. Construct the RC bandpass filter circuit shown in Figure 1-4a.
2. Calculate the low-frequency break point using the following formula.

$$F_L = \frac{1}{2\pi RC}$$

   where F_L = low-frequency break point
   $R = R_1 + R_2$
   $C = C_1$

3. Calculate the high-frequency break point using the following formula.

$$F_H = \frac{1}{2\pi RC}$$

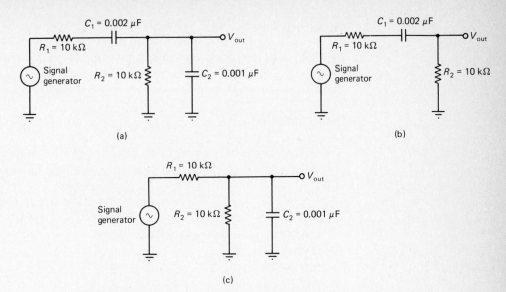

FIGURE 1-4 Passive bandpass filter, (a) Schematic diagram. (b) Low frequency equivalent circuit. (c) High frequency equivalent circuit.

where F_H = high-frequency break point

$$R = Rth = \frac{R_1 \times R_2}{R_1 + R_2}$$

$$C = C_2$$

4. Measure the actual upper and lower break frequencies by setting the signal generator output voltage to 4 V p–p and varying its output frequency from 100 Hz to 100 kHz.

5. Calculate the percentage error between the calculated and actual upper and lower break frequencies.

6. Plot the frequency response curve for the bandpass filter on semilog graph paper.

7. Measure the insertion loss of the filter.

8. Design a bandpass filter with a lower cutoff frequency F_L = 1.7 kHz and an upper cutoff frequency F_U = 16 kHz. Use resistances R_1 = 10k ohms and R_2 = 10k ohms.

9. Construct the bandpass filter circuit designed in step 8, and plot the frequency response curve on semilog graph paper.

SECTION D   Bandstop Filter

In this section, the frequency response of a passive RC bandstop filter is examined. A bandstop filter, like a bandpass filter, is the combination of a lowpass and a highpass filter. The schematic diagram for the bandstop filter circuit used in this section is shown in Figure 1-5a. The high-frequency equivalent circuit is shown in Figure 1-5b. C_1 and C_2 look like short circuits to high frequencies; thus, high frequencies are passed directly to the output. The low-frequency equivalent circuit is shown in Figure 1-5c. C_1, C_2, and C_3 look like open circuits to low frequencies; thus, low frequencies are passed directly to the output through R_1 and R_2. However, a band of intermediate frequencies are attenuated. The equivalent circuit for frequencies near the upper break frequency is shown in Figure 1-5d. The input signal voltage divides between R_1 and C_3. The upper cutoff frequency occurs at a frequency F_U = $1/2\pi R_1 C_3$. The equivalent circuit for frequencies near the lower break frequency is shown in Figure 1-5e. The input signal voltage divides between R_3 and C_1. The lower cutoff frequency occurs at a frequency $F_L = 1/2\pi R_3 C_1$. The values for R_1, R_3, C_1, and C_3 are selected such that the two break frequencies are separated by two octaves. Therefore, the minimum output voltage occurs at a frequency an octave above the lower break frequency and an octave below the upper break frequency.

Procedure

1. Construct the RC bandstop filter circuit shown in Figure 1-5a.
2. Calculate the low-frequency break point using the following formula.

$$F_L = \frac{1}{2\pi RC}$$

where F_L = low-frequency break point
$R = R_1$
$C = C_3$

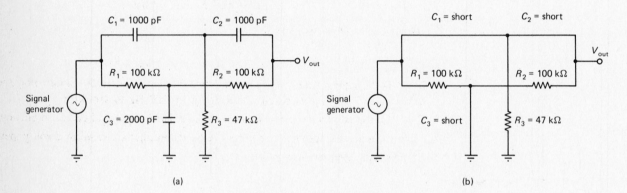

(a)   (b)

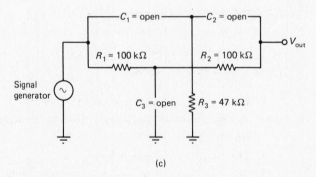

(c)

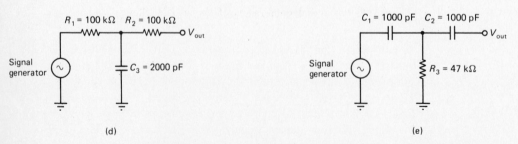

(d)   (e)

FIGURE 1-5 Passive bandstop filter, (a) Schematic diagram. (b) High frequency equivalent circuit. (c) Low frequency equivalent circuit. (d) Equivalent circuit for frequencies near the upper break frequency. (e) Equivalent circuit for frequencies near the lower break frequency.

3. Calculate the high-frequency break point using the following formula.

$$F_H = \frac{1}{2\pi RC}$$

where F_H = high-frequency break point
$R = R_3$
$C = C_1$

4. Measure the actual upper and lower break frequencies by setting the signal generator output voltage to 4 V p–p and varying its output frequency from 100 Hz to 100 kHz.

5. Calculate the percentage error between the calculated and actual upper and lower break frequencies.

6. Calculate the center frequency for the filter (i.e., the frequency with the lowest output amplitude) using the following formula.

$$F_c = \frac{1}{2\pi\sqrt{R_1 C_3 R_3 C_1}} = \frac{1}{2\pi RC}$$

where F_c = center frequency
$R = R_1 = R_2$
$C = C_1 = C_2$

7. Measure the actual center frequency by setting the signal generator output voltage to 4 V p–p and varying its output frequency from 100 Hz to 100 kHz.

8. Calculate the percentage error between the calculated and actual center frequency.

9. Plot the frequency response curve for the bandstop filter on semilog graph paper.

10. Measure the insertion loss of the filter.

SECTION E   Summary

Write a brief summary of the concepts presented in this experiment on passive RC filters. Include the following items:

1. The characteristics of passive RC filters.
2. The frequency response characteristics of lowpass, highpass, bandpass, and bandstop filters.
3. The concept of insertion loss.
4. The concept of upper and lower cutoff frequencies.
5. The roll-off characteristics of single-pole filters.

EXPERIMENT 1   ANSWER SHEET

NAME: _____ CLASS: _____ DATE: _____

SECTION A

2. F_U (calculated) = _____   3. F_U (measured) = _____

4. % error = _____   6. IL = _____

7. C = _____

SECTION B

2. F_L (calculated) = _____   3. F_L (measured) = _____

4. % error = _____   6. IL = _____

7. C = _____

SECTION C

2. F_L (calculated) = _____   3. F_H (calculated) = _____

4. F_L (measured) = _____   5. % error (F_L) = _____

     F_U (measured) = _____   % error (F_H) = _____

7. IL = _____   8. C_1 = _____

                                   C_2 = _____

SECTION D

2. F_L (calculated) = _____   3. F_H (calculated) = _____

4. F_L (measured) = _____   5. % error (F_L) = _____

     F_H (measured) = _____   % error (F_H) = _____

6. F_c (calculated) = _____   7. F_c (measured) = _____

8. % error = _____   10. IL = _____

Experiment 2

PASSIVE LC FILTERS

REFERENCE TEXT: James Harter and Paul Lin, *Essentials of Electric Circuits*, 2nd edition (Englewood Cliffs, N.J.: Prentice-Hall, 1986).

1. Chapter 26, Resonance.
2. Chapter 29, Introduction to Filters.

OBJECTIVES

1. To observe the frequency response characteristics of passive LC filters.
2. To observe the frequency response characteristics of bandpass and bandstop filters.

INTRODUCTION

LC filters are tuned circuits and, like their RC counterparts, are passive frequency dependent networks that are often used in electronic communications circuits to separate a single frequency or a specific band of frequencies from a composite waveform. LC networks can be configured as either bandpass or bandstop filters. In this exercise bandpass and bandstop filters are constructed using inductors (coils), resistors, and capacitors, and then their frequency response characteristics are measured. LC filters typically have sharper roll-off characteristics and narrower bandwidths than their RC counterparts. The roll-off and bandwidth characteristics depend on the quality (Q) factor of the coil and the resonant frequency of the tuned circuit. The resonant frequency of an LC tuned circuit is the frequency at which the inductor and capacitor give up and take on energy at the same rate. The resonant frequency is also the frequency that falls in the center of a tuned circuit's response curve. The Q factor is the ratio of a coil's reactive power to its real power (i.e., $Q = I^2X_L/I^2R = X_L/R$). The dc resistance of an ideal coil is 0 ohm, and consequently, an ideal coil dissipates no power and has an infinite Q. However, in a more practical circuit, power is dissipated in the dc resistance of a coil, and any series or shunt resistances add to the coil resistance, decreasing the Q and widening the bandwidth. The block diagram for the test circuit used in this experiment is shown in Figure 2-1.

9

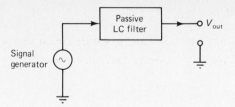

FIGURE 2-1 Passive LC filter test circuit

MATERIALS REQUIRED

Equipment:

1 — protoboard

1 — medium-frequency signal generator (1 MHz)

1 — standard oscilloscope (10 MHz)

1 — assortment of test leads and hookup wire

Parts List:

1 — 3 mH inductor

1 — 22 ohm resistor

1 — 1 k-ohm resistor

1 — 10 k-ohm resistor

1 — 1 M-ohm resistor

1 — 0.001 μF capacitor

1 — 0.01 μF capacitor

1 — 0.05 μF capacitor

SECTION A   Series Resonant LC Bandpass Filter

In this section the frequency response characteristics for a series resonant LC bandpass filter are examined. The impedance of a series tuned circuit is minimum at resonance (i.e., at the resonant frequency). At resonance the reactance of the capacitor and the inductor are equal, and thus, they cancel each other. Consequently, the total series impedance $Z_s = R_{dc} + jX_L - jX_C = R_{dc}$, where R_{dc} is the dc resistance of the inductor. The series impedance increases as frequency increases or decreases from resonance. For Q factors greater than 10, the bandwidth $B = F_r/Q$ (where F_r is the resonant frequency). The schematic diagram for the bandpass filter circuit used in this section is shown in Figure 2-2. At resonance, Z_s is approximately equal to R_{dc}, and a large portion of the input signal is dropped across R_1. At frequencies above or below resonance, Z_s increases and the output voltage decreases. The total circuit Q is the inductive reactance divided by the total circuit resistance (i.e., $Q = X_L/[R_{dc} + R_1]$). R_1 adds to the series resistance of the coil, thus reducing the overall Q and increasing the bandwidth of the filter.

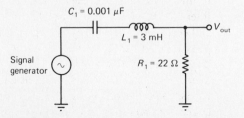

FIGURE 2-2 Series resonant LC bandpass filter

Procedure

1. Construct the LC bandpass filter circuit shown in Figure 2-2.
2. Calculate the resonant frequency of the LC network using the following formula.

$$F_r = \frac{1}{2\pi\sqrt{LC}}$$

where F_r = resonant frequency
  $C = C_1$
  $L = L_1$

3. Measure the actual resonant frequency by setting the signal generator output voltage to 4 V p–p and varying its output frequency from 50 to 150 kHz.
4. Calculate the percentage error between the calculated and actual resonant frequencies using the following formula.

$$\% \text{ error} = \frac{\text{Measured value} - \text{calculated value}}{\text{calculated value}} \times 100$$

5. Plot the frequency response curve for the bandpass filter on semilog graph paper.
6. From the frequency response curve, determine the bandwidth of the filter.
7. Calculate the total circuit Q using the following formula.

$$Q = \frac{F_r}{B}$$

where Q = circuit Q factor
  F_r = resonant frequency
  B = filter bandwidth

8. Measure the insertion loss for the filter at resonance using the following formula.

$$IL = \frac{V_{\text{out}}}{V_{\text{in}}}$$

where IL = insertion loss
  V_{out} = filter output voltage
  V_{in} = filter input voltage

9. Design a series resonant bandpass filter with a resonant frequency $F_r = 650$ kHz. Use an inductance $L = 3$ mH.
10. Construct the bandpass filter circuit designed in step 9, and plot its frequency response curve on semilog graph paper.

SECTION B   Parallel Resonant LC Bandpass Filter

In this section the frequency response characteristics of a parallel resonant LC bandpass filter are examined. The impedance of a parallel tuned circuit is maximum at resonance and decreases as frequency increases or decreases. For Q factors greater than 10, the impedance of a parallel tuned circuit, Z_p, is approximately equal to QX_L. The schematic diagram for the bandpass filter circuit used in this section is shown in Figure 2-3. At resonance, Z_p is maximum and a significant portion of the input signal is dropped across the LC network and appears at V_{out}. For input frequencies above or below the resonant frequency, Z_p decreases and a larger portion of the input signal is dropped across R_1. Consequently, V_{out} decreases.

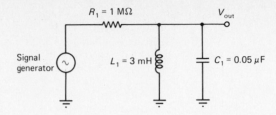

FIGURE 2-3   Parallel resonant LC bandpass filter

Procedure

1. Construct the LC bandpass filter circuit shown in Figure 2-3.
2. Calculate the resonant frequency of the LC network.
3. Measure the actual resonant frequency by setting the signal generator output voltage to 10 V p–p and varying its output frequency from 10 to 100 kHz.
4. Calculate the percentage error between the calculated and actual resonant frequencies.
5. Plot the frequency response curve for the bandpass filter on semilog graph paper.
6. From the frequency response curve, determine the bandwidth of the filter.
7. Calculate the total circuit Q.
8. Measure the insertion loss for the filter at resonance.
9. Design a parallel resonant bandpass filter with a resonant frequency $F_r = 9.2$ kHz. Use an inductance $L = 3$ mH.
10. Construct the bandpass filter circuit designed in step 9, and plot its frequency response curve on semilog graph paper.

SECTION C   Series Resonant LC Bandstop Filter

In this section the frequency response characteristics of a series resonant LC bandstop filter are examined. With bandstop filters, the break frequency is the frequency at which the amplitude of the output signal decreases by some specified amount. The bandwidth for a bandstop filter is generally given as either 3-dB or 10-dB rejection bandwidths. The schematic diagram for the bandstop filter circuit used in this section is shown in Figure 2-4. At resonance, Z_s is minimum. Consequently, most of the input voltage is dropped across R_1 and V_{out} is minimum. For input frequencies above or below resonance, Z_s increases, causing a proportional increase in V_{out}.

Procedure

1. Construct the LC bandstop filter circuit shown in Figure 2-4.
2. Calculate the resonant frequency of the LC network.
3. Measure the actual resonant frequency by setting the signal generator output voltage to 10 V p–p and varying its output frequency from 5 to 500 kHz.
4. Calculate the percentage error between the calculated and actual resonant frequencies.
5. Plot the frequency response curve for the bandstop filter on semilog graph paper.
6. From the frequency response curve, determine the 3-dB and 10-dB rejection bandwidths.
7. Calculate the total circuit Q.

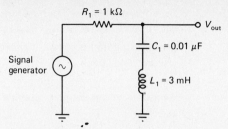

FIGURE 2-4   Series resonant LC bandstop filter

8. Measure the insertion loss and the loss at resonance.

9. Replace the 1 k-ohm resistor with a 10 k-ohm resistor and repeat steps 5 through 8.

10. Describe the results observed when the 1 k-ohm resistor is replaced by the 10 k-ohm resistor. Include the effects on circuit Q and bandwidth.

11. Design a series resonant bandstop filter with a resonant frequency $F_r = 13$ kHz. Use an inductance $L = 3$ mH.

12. Construct the bandstop filter circuit designed in step 11, and plot its frequency response curve on semilog graph paper.

Section D Parallel Resonant Bandstop Filter

In this section the frequency response characteristics of a parallel resonant LC bandstop filter are examined. The schematic diagram for the bandstop filter circuit used in this section is shown in Figure 2-5. At resonance, Z_p is maximum. Consequently, most of the input voltage is dropped across the LC network and the output voltage is minimum. For input frequencies above and below resonance, Z_p decreases, causing a proportional increase in V_{out}.

Procedure

1. Construct the LC bandstop filter circuit shown in Figure 2-5.

2. Calculate the resonant frequency of the LC network.

3. Measure the actual resonant frequency by setting the signal generator output voltage to 10 V p–p and varying its output frequency from 50 to 150 kHz.

4. Calculate the percentage error between the calculated and actual resonant frequencies.

5. Plot the frequency response curve for the bandstop filter on semilog graph paper.

6. From the frequency response curve, determine the 3-dB and 10-dB rejection bandwidths.

7. Calculate the total circuit Q.

8. Measure the insertion loss and the loss at resonance.

9. Place a 10 k-ohm resistor in parallel with the LC tank circuit and repeat steps 5 through 8.

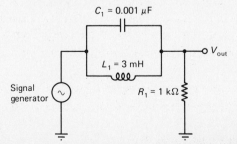

FIGURE 2-5   Parallel resonant LC bandstop filter

10. Describe the results observed when the 10 k-ohm resistor is placed across the LC tank circuit. Include the effects on circuit Q and bandwidth.

11. Design a parallel resonant bandstop filter with a resonant frequency $F_r = 65$ kHz. Use an inductance $L = 3$ mH.

12. Construct the bandstop filter circuit designed in step 11, and plot its frequency response curve on semilog graph paper.

SECTION E   Summary

Write a brief summary of the concepts presented in this experiment on passive LC filters. Include the following items:

1. The characteristics of series and parallel tuned circuits.
2. The characteristics of bandpass and bandstop filters.
3. The concept of insertion loss.
4. The effects of loading a tank circuit.

EXPERIMENT 2   Answer Sheet

NAME: _____ CLASS: _____ DATE: _____

SECTION A

2. F_r (calculated) = _____   3. F_r (measured) = _____

4. % error = _____   6. B = _____

7. Q = _____   8. IL = _____

9. C = _____

SECTION B

2. F_r (calculated) = _____   3. F_r (measured) = _____

4. % error = _____   6. B = _____

7. Q = _____   8. IL = _____

9. C = _____

SECTION C

2. F_r (calculated) = _____   3. F_r (measured) = _____

4. % error = _____   6. B (-3 dB) = _____

7. Q = _____ B (-10 dB) = _____

9. B (-3 dB) = _____   8. IL = _____

   B (-10 dB) = _____

   Q = _____

   IL = _____

10. _____

   _____

   _____

11. C = _____

SECTION D

2. F_r (calculated) = _____   3. F_r (measured) = _____

4. % error = _____   6. B (-3 dB) = _____

7. $Q =$ _____ $B\ (-10\ \text{dB}) =$ _____

9. $B\ (-3\ \text{dB}) =$ _____   8. $IL =$ _____

   $B\ (-10\ \text{dB}) =$ _____

   $Q =$ _____

   $IL =$ _____

10. _____

    _____

    _____

11. $C =$ _____

ACTIVE FILTERS

REFERENCE TEXT: Robert F. Doughlin and Frederick F. Driscoll, *Operational Amplifiers and Lienar Integrated Circuits,* 3rd edition (Englewood Cliffs, N.J.: Prentice-Hall, 1987).

Chapter 11, Active Filters.

OBJECTIVES

1. To observe the gain-versus-frequency response characteristics of active filters.
2. To observe the operation of op amp filters.
3. To observe the frequency response characteristics of unity-gain lowpass and high-pass filters.

INTRODUCTION

Active filters have frequency response characteristics similar to those of passive filters, except with gain. Therefore, the insertion loss is 0 dB. In this exercise unity-gain active filters are constructed using resistors, capacitors, and integrated circuit operational amplifiers. A lowpass filter is a circuit that has a constant output voltage from 0 Hz (dc) up to the filter break frequency (sometimes called the upper cutoff frequency). As the frequency increases above the break frequency, the output voltage decreases. A highpass filter is a circuit that has a constant output voltage for all frequencies above the break frequency (the break frequency for a highpass filter is sometimes called the lower cutoff frequency). A bandpass filter has a constant output voltage for frequencies within its passband and has an upper and a lower cutoff frequency. Frequencies below the lower cutoff frequency or above the upper cutoff frequency are attenuated. A bandstop filter rejects frequencies with a specific band and passes all others. Frequencies above the lower cutoff frequency or below the upper cutoff frequency are attenuated. The block diagrams for the active filter circuits used in this experiment are shown in Figure 3-1. Figure 3-1a shows a simple single-pole active filter, which has a −20 dB/decade roll-off. Figure 3-1b shows a two-pole unity-gain Salen-Key filter, which has a −40 dB/decade roll-off. Either configura-

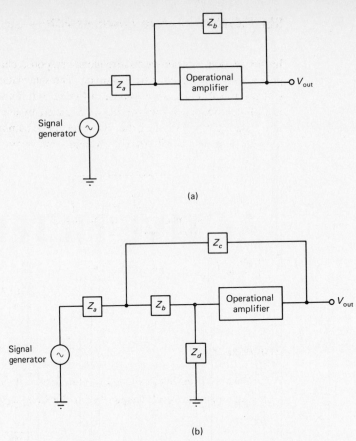

(a)

(b)

FIGURE 3-1   Active filters, (a) Single-pole, (b) Two-pole.

tion can be used as a lowpass or a highpass filter, depending on the choice and placement of components. Also, bandpass and bandstop filters can be configured by combining a lowpass and a highpass filter.

MATERIALS REQUIRED

Equipment:

1 — protoboard
1 — dual dc power supply (+ 15 V dc and − 15 V dc)
1 — medium-frequency signal generator (1 MHz)
1 — standard oscilloscope (10 MHz)
1 — assortment of test leads and hookup wire

Parts List:

1 — operational amplifier (741C or equivalent)
2 — 1 k-ohm resistors
1 — 10 k-ohm resistor
1 — 22 k-ohm resistor
1 — 0.001 μF capacitor
2 — 0.01 μF capacitors

SECTION A Active Lowpass Filter—single-pole, noninverting

In this section the gain-versus-frequency response characteristics of a unity-gain single-pole active lowpass filter are examined. The schematic diagram for the lowpass filter circuit used in this section is shown in Figure 3-2. This filter produces a -20 dB/decade roll-off for frequencies above the upper cutoff frequency. At high frequencies $C1$ looks like a short circuit, shorting the signal generator output to ground. At low frequencies $C1$ is open, and the circuit becomes a unity-gain noninverting voltage follower.

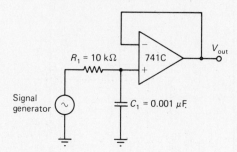

FIGURE 3-2 Active lowpass filter—single-pole, noninverting configuration.

Procedure

1. Construct the lowpass filter circuit shown in Figure 3-2.
2. Calculate the break frequency for the filter using the following formula.

$$F_U = \frac{1}{2\pi RC}$$

where F_U = upper cutoff frequency
$R = R_1$
$C = C_1$

3. Measure the actual break frequency by setting the amplitude of the signal generator output voltage to 1 V p–p and varying its frequency from 0 Hz to 10 kHz.
4. Calculate the percentage error between the calculated and actual break frequencies using the following formula.

$$\% \text{ error} = \frac{\text{measured value} - \text{calculated value}}{\text{calculated value}} \times 100$$

5. Plot the gain-versus-frequency response curve for the lowpass filter on semilog graph paper.
6. Measure the gain of the filter using the following formula.

$$A_v = \frac{V_{out}}{V_{in}}$$

where A_v = voltage gain
V_{out} = op amp output voltage
V_{in} = op amp input voltage

7. Design an active single-pole lowpass filter with an upper cutoff frequency $F_U =$ 4.8 kHz. Use the noninverting configuration shown in Figure 3-2 and a capacitance $C_1 = 0.001$ μF.
8. Construct the lowpass filter circuit designed in step 7, and plot its gain-versus-frequency response curve on semilog graph paper.

SECTION B   Active Lowpass Filter—single-pole, inverting

In this section the gain-versus-frequency response characteristics of a unity-gain single-pole active lowpass filter are examined. The schematic diagram for the lowpass filter circuit used in this section is shown in Figure 3-3. This filter produces a -20 dB/decade roll-off for frequencies above the break frequency. At high frequencies C_1 is a short circuit, shunting the feedback resistor and reducing the circuit gain to 0. At low frequencies C_1 is open, and the gain of the inverting amplifier is $-R_2/R_1$ (i.e., unity).

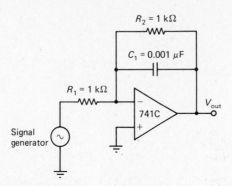

FIGURE 3-3   Active lowpass filter—single-pole, inverting configuration.

Procedure

1. Construct the lowpass filter circuit shown in Figure 3-3.
2. Calculate the break frequency for the filter using the following formula.

$$F_U = \frac{1}{2\pi RC}$$

where F_U = upper cutoff frequency
$\qquad R = R_2$
$\qquad C = C_1$

3. Measure the actual break frequency by setting the amplitude of the signal generator output voltage to 1 V p–p and varying its frequency from 20 to 200 kHz.
4. Calculate the percentage error between the calculated and actual break frequencies.
5. Plot the gain-versus-frequency response curve for the lowpass filter on semilog graph paper.
6. Measure the gain of the filter.
7. Design a unity-gain single-pole active lowpass filter with an upper cutoff frequency $F_U = 16$ kHz. Use an inverting configuration and a capacitance $C_1 = 0.001$ µF.

SECTION C   Active Lowpass Filter—two-pole, noninverting

In this section the gain-versus-frequency response characteristics of a unity-gain two-pole active lowpass filter are examined. The schematic diagram for the lowpass filter circuit used in this section is shown in Figure 3-4. This filter is a two-pole Salen-Key unity-gain filter that produces a -40 dB/decade roll-off for frequencies above the upper cutoff frequency. For high frequencies C_2 is a short circuit, placing the noninverting input of the op amp at ac ground. For low frequencies C_1 and C_2 are open, and the circuit is a unity-gain voltage follower.

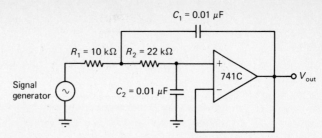

FIGURE 3-4 Active lowpass filter—two-pole, noninverting configuration.

Procedure

1. Construct the lowpass filter circuit shown in Figure 3-4.
2. Calculate the break frequency for the filter using the following formula.

$$F_U = \frac{1}{2\pi\sqrt{R_1 R_2 C_1 C_2}}$$

where F_U = resonant frequency

3. Measure the actual break frequency by setting the amplitude of the signal generator output voltage to 1 V p–p and varying its frequency from 0 Hz to 10 kHz.
4. Calculate the percentage error between the calculated and actual break frequencies.
5. Plot the gain-versus-frequency response curve for the lowpass filter on semilog graph paper.
6. Measure the gain of the filter.

SECTION D   Active Highpass Filter—single-pole, noninverting

In this section the gain-versus-frequency response characteristics of a unity-gain single-pole highpass filter are examined. The schematic diagram for the highpass filter circuit used in this experiment is shown in Figure 3-5. This filter produces a -20 dB/decade roll-off for frequencies below the break frequency. At low frequencies C_1 is open, which prevents the signal generator's output from reaching the noninverting input to the op amp, thus reducing the filter output to 0. At high frequencies C_1 is a short, and the circuit is reduced to a unity gain voltage follower.

Procedure

1. Construct the highpass filter circuit shown in Figure 3-5.
2. Calculate the break frequency for the filter using the following formula.

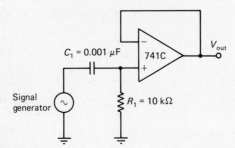

FIGURE 3-5 Active highpass filter—single-pole, noninverting configuration.

$$F_L = \frac{1}{2\pi RC}$$

where F_L = lower cutoff frequency
$R = R_1$
$C = C_1$

3. Measure the actual break frequency by setting the amplitude of the signal generator output voltage to 1 V p–p and varying its frequency from 0 Hz to 10 kHz.
4. Calculate the percentage error between the calculated and actual break frequencies.
5. Plot the gain-versus-frequency response curve for the highpass filter on semilog graph paper.
6. Measure the gain of the filter.
7. Design a unity-gain single-pole active highpass filter with a lower cutoff frequency F_L = 159 kHz. Use the noninverting configuration shown in Figure 3-5 and a capacitance C_1 = 0.001 μF.

SECTION E Active Highpass Filter—two-pole, noninverting

In this section the gain-versus-frequency response characteristics of a unity gain two-pole active highpass filter are examined. The schematic diagram for the highpass filter circuit used in this section is shown in Figure 3-6. This filter produces a −40 dB/decade roll-off for frequencies below the break frequency. For low frequencies C_1 and C_2 are opens, thus blocking the signal generator's output signal. At high frequencies C_1 and C_2 are shorts, and the circuit is a unity-gain noninverting amplifier.

Procedure

1. Construct the highpass filter circuit shown in Figure 3-6.
2. Calculate the break frequency using the following formula.

$$F_L = \frac{1}{2\pi\sqrt{R_1 R_2 C_1 C_2}}$$

where F_L = resonant frequency

3. Measure the actual break frequency by setting the amplitude of the signal generator output voltage to 1 V p–p and varying its frequency from 0 Hz to 10 kHz.
4. Calculate the percentage error between the calculated and actual break frequencies.
5. Plot the gain-versus-frequency response curve for the highpass filter on semilog graph paper.
6. Measure the gain of the filter.

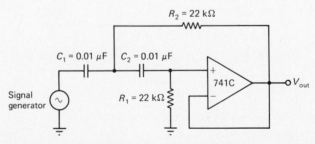

FIGURE 3-6 Active highpass filter—two-pole, noninverting configuration.

SECTION F   Summary

Write a brief summary of the concepts presented in this experiment on active filters. Include the following items:

1. The characteristics of active filters.
2. The gain-versus-frequency response characteristics of lowpass, highpass, bandpass, and bandstop filters.
3. The concept of upper and lower cutoff frequencies.
4. The roll-off characteristics of single- and two-pole filters.
5. The concept of unity-gain filters.

EXPERIMENT 3   Answer Sheet

NAME: _____ CLASS: _____ DATE: _____

SECTION A

2. F_U (calculated) = _____   3. F_U (measured) = _____

4. % error = _____   6. Av = _____

7. R = _____

SECTION B

2. F_U (calculated) = _____   3. F_U (measured) = _____

4. % error = _____   6. Av = _____

7. R = _____

SECTION C

2. F_U (calculated) = _____   3. F_U (measured) = _____

4. % error = _____   6. Av = _____

SECTION D

2. F_L (calculated) = _____   3. F_L (measured) = _____

4. % error = _____   6. Av = _____

7. R = _____

SECTION E

2. F_r (calculated) = _____   3. F_r (measured) = _____

4. % error = _____   6. Av = _____

SIGNAL ANALYSIS

REFERENCE TEXT: *Fundamentals of Electronic Communications Systems.*
Chapter 1, Signal Analysis.

OBJECTIVES

1. To observe the frequency spectrum for nonsinusoidal periodic (complex) waveforms.
2. To observe complex waveforms with half-wave symmetry.
3. To observe the effects of bandlimiting complex repetitive waveforms.

INTRODUCTION

In electronic communications circuits it is often necessary to analyze and predict the performance of a circuit based on the voltage, bandwidth, and frequency composition of the information signal. In essence, signal analysis is the mathematical analysis of the frequency, bandwidth, and voltage level of a signal. Electrical signals are voltage- or current-time variations that can be represented by a series of sine or cosine waves. Essentially, any repetitive waveform that consists of more than one sine or cosine wave is a nonsinusoidal or complex wave. A mathematical series developed by Baron Jean Fourier (appropriately called the Fourier series) is used to mathematically analyze a complex repetitive waveform. The Fourier series states that all complex periodic waves consist of an average dc component and a series of harmonically related sine and cosine waves. The fundamental frequency is the minimum frequency necessary to represent a waveform and is also the frequency or repetition rate of the waveform. In this experiment several complex waveforms are examined in the time domain using a standard oscilloscope and passive filters. The block diagram for the test circuit used in this experiment is shown in Figure 4-1.

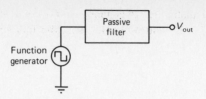

FIGURE 4-1 Signal analysis test circuit

MATERIALS REQUIRED

Equipment:

1 — protoboard
1 — medium-frequency function generator (1 MHz)
1 — standard oscilloscope
1 — capacitance substitution box (0.0001 μF to 1 μF)
1 — assortment of test leads and hookup wire

Parts List:

1 — 3-mH inductor
1 — 20-ohm resistor
1 — 100-ohm resistor
1 — 1k-ohm resistor
1 — 10k-ohm resistor
1 — 100k-ohm resistor
1 — 1M-ohm resistor
1 — 0.001 μF capacitor

SECTION A   Harmonic Composition of a Square Wave

In this section the harmonic composition of a square wave is examined. A square wave is a complex wave that exhibits half-wave symmetry. That is, a square wave is a nonsinusoidal repetitive wave where the even harmonics in the series for both the sine and cosine terms are 0. Thus, a square wave comprises a fundamental frequency and a series of odd harmonics. The schematic diagram for the test circuit used in this section is shown in Figure 4-2. A square wave output signal from the function generator is passed through an LC bandpass filter which is tuned first to the fundamental frequency of the square wave, then to the third, fifth, and seventh harmonics, respectively. The amplitudes of the first four odd harmonics are measured and then compared to their predicted values.

Procedure

1. Construct the test circuit shown in Figure 4-2.

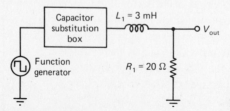

FIGURE 4-2 Harmonic composition test circuit

2. Calculate the capacitance of C_1 for resonant frequencies of 20 kHz, 60 kHz, 100 kHz, and 140 kHz using the following formula.

$$C = \frac{\left(\frac{1}{2\pi F}\right)^2}{L_1}$$

where C = capacitance C_1
F = 20 kHz, 60 kHz, 100 kHz, and 140 kHz
L = L_1 = 3 mH

3. Set the output of the function generator to a 4 V p–p square wave and a repetition rate equal to 20 kHz.

4. Calculate the peak voltages for the first four odd harmonics (i.e., first, third, fifth, and seventh) of the 20 kHz square wave using the following formula.

$$V_n = \frac{4V}{n\pi}$$

where V_n = peak voltage of the nth harmonic
V = peak amplitude of the square wave (2 V)
n = 1, 3, 5, and 7

5. Using the peak voltages calculated in step 4, draw the frequency spectrum for the 20 kHz square wave.

6. Measure the peak amplitudes of the first four odd harmonics of the 20 kHz square wave at the output of the filter as follows. Set the capacitance of the substitution box to the value determined in step 2 for a frequency of 20 kHz and measure the peak voltage at the output of the filter. In turn, set the capacitance of the substitution box to the remaining three values of C_1 determined in step 2 for 60 kHz, 100 kHz, and 140 kHz, and measure the peak voltage of the third, fifth, and 7th harmonics of the 20 kHz square wave, respectively. (The capacitance of the substitution box may have to be varied slightly from the calculated values.)

7. Using the peak voltages measured in step 6, draw the frequency spectrum for the 20 kHz square wave.

8. Compare the voltages measured in step 6 and the frequency spectrum drawn in step 7 with the voltages calculated in step 4 and the frequency spectrum drawn in step 5. Because of the insertion loss of the bandpass filter, the voltages measured in step 6 should all be lower than those calculated in step 4 by a proportionate amount. The exact values of the voltages are not as important as their ratios (i.e., the ratio of the calculated to the measured voltage for 20 kHz should equal the ratio of the calculated to the measured voltages for 60 kHz, 100 kHz, and 140 kHz). This ratio is the insertion loss ratio for the bandpass filter. Determine the insertion loss ratio for the bandpass filter.

9. Measure the insertion loss ratio of the bandpass filter by setting the function generator to a 4-V-p–p sine wave and a frequency of 20 kHz. Set the capacitance of the substitution box to the value determined in step 2 for 20 kHz. Calculate the insertion-loss ratio using the following formula.

$$ILR = \frac{V_{out}}{V_{in}}$$

where ILR = insertion loss ratio
V_{out} = peak voltage at the output of the filter
V_{in} = 2 V p

SECTION B   The Effect of Band Limiting Complex Waves

In this section the effect of band limiting signals is examined. Every communications system has a limited bandwidth and, therefore, has a limiting effect on signals that are propagated through it. If a communications channel is considered as an ideal linear phase lowpass filter with a finite bandwidth, the frequency spectrum for nonsinusoidal repetitive waveforms that pass through the channel is changed. The harmonic frequency components that are higher in frequency than the upper cutoff frequency for the filter are attenuated or removed. Consequently, the shape of the waveform is changed. In this section a triangular wave is passed through a single-pole passive RC lowpass filter. The upper cutoff frequency of the filter is reduced in successive steps. After each bandwidth reduction, the output waveform is observed with an oscilloscope. A triangular wave has half-wave symmetry (i.e., it consists of odd harmonics only). The schematic diagram for the test circuit used in this section is shown in Figure 4-3. The RC network comprising R_1 and C_1 is a lowpass filter.

Procedure

1. Construct the test circuit shown in Figure 4-3.
2. Calculate the upper cutoff frequency for the lowpass filter circuit shown in Figure 4-3 for a capacitance $C_1 = 0.001$ μF and resistance values $R_1 = 100$, 1k, 10k, 33k, 100k, and 1 Mohms using the following formula.

$$F_U = \frac{1}{2\pi RC}$$

where F_u = upper cutoff frequency
  $C = C_1 = 0.001$ μF
  $R = R_1$ (100, 1k, 10k, 33k, 100k, and 1 Mohm)

3. Set the output of the function generator to a 10 V-p–p triangular wave and a repetition rate equal to 20 kHz.
4. Calculate the peak voltages for the first five odd harmonics (first, third, fifth, seventh, and ninth) of the 20 kHz triangular wave using the following formula.

$$V_n = \frac{8\ V}{(n\pi)^2}$$

where V_n = peak amplitude of the nth harmonic
  V = peak amplitude of the triangular wave (5 V)
  n = 1, 3, 5, 7, and 9

5. Using the peak voltages calculated in step 4, draw the frequency spectrum for the 20 kHz triangular wave.
6. Construct a table listing the six upper cutoff frequencies for the lowpass filter calculated in step 2. For each resistance value, list which of the first five harmonic frequencies calculated in step 4 will pass through the filter.

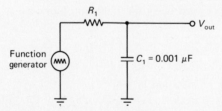

FIGURE 4-3   Test circuit for analyzing the effects of band-limiting complex waves.

7. Insert each of the six resistance values listed in step 2 for R_1, and for each value of R_1, sketch the filter output waveform.

8. Describe the waveforms sketched in step 7 in terms of amplitude and frequency content.

SECTION C   Summary

Write a brief summary of the concepts presented in this experiment on signal analysis. Include the following items:

1. The harmonic composition of a nonsinusoidal complex wave.
2. The harmonic composition of a wave that exhibits half-wave symmetry.
3. The harmonic composition of both a square wave and a triangular wave.
4. The effects of band limiting a complex repetitive wave in terms of waveshape and frequency composition.

EXPERIMENT 4   Answer Sheet

NAME: _____ CLASS: _____ DATE: _____

SECTION A

2. C (20 kHz) = _____ C (60 kHz) = _____

   C (100 kHz) = _____ C (140 kHz) = _____

4. V_1 = _____ V_3 = _____

   V_5 = _____ V_7 = _____

5.

Voltage / Frequency

6. V_1 = _____ V_3 = _____

   V_5 = _____ V_7 = _____

7

Voltage / Frequency

8. _____

_____

_____

9. ILR = _____

SECTION B

2. F_U (100 Ω) = _____ F_U (1 kΩ) = _____

   F_U (10k Ω) = _____ F_U (33 kΩ) = _____

   F_U (100 kΩ) = _____ F_U (1 MΩ) = _____

4. V_1 = _____ V_3 = _____

   V_5 = _____ V_7 = _____

   V_9 = _____

5.

Voltage / Frequency

6.

R (kΩ)	F	Harmonics
0.1		
1		
10		
33		
100		
1000		

7.

Vertical sensitivity _____ V/cm

Time base _____ sec/cm

Vertical sensitivity _____ V/cm

Time base _____ sec/cm

Vertical sensitivity _____ V/cm

Time base _____ sec/cm

Vertical sensitivity _____ V/cm

Time base _____ sec/cm

Vertical sensitivity _____ V/cm

Time base _____ sec/cm

Vertical sensitivity _____ V/cm

Time base _____ sec/cm

8. _____

_____

_____

_____

Experiment 5

LINEAR SUMMING

REFERENCE TEXT: *Fundamentals of Electronic Communications Systems.*
Chapter 3, Linear Summing.

OBJECTIVES

1. To observe linear summing of two sine waves.
2. To observe linear summing of a sine wave and a square wave.

INTRODUCTION

Linear summing (or linear mixing as it's sometimes called) occurs whenever two or more signals are combined in a linear circuit. When two signals are linearly combined, the resulting waveform is the algebraic sum of the two signals; no harmonics or cross-product frequencies are produced. The frequency spectrum at the output of a linear device contains only the original input frequencies. In this exercise two signals are combined in an op amp summer, which is a linear circuit. The two signals are derived from the same source to ensure stable triggering of the oscilloscope and to enable the observer to easily predict the shape and frequency content of the output waveform. The block diagram for the test circuit used in this experiment is shown in Figure 5-1. The wave shaping network sets the amplitude and phase of V_1 and V_2, and the linear summer algebraically combines them.

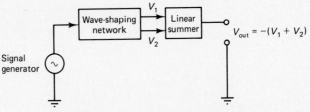

FIGURE 5-1 Linear summer block diagram

33

MATERIALS REQUIRED

Equipment:

1 — protoboard
1 — dual dc power supply ($+15$ V dc and -15 V dc)
1 — audio signal generator
1 — standard oscilloscope (10 MHz)
1 — assortment of test leads and hookup wire

Parts List:

3 — operational amplifiers (741C or equivalent)
5 — equal-value resistors (1 to 10 k ohms)
1 — 10 k-ohm variable resistor

SECTION A   Linear Summing of Two Sine Waves

In this section two sine waves of equal frequency (V_1 and V_2) are combined, first in phase and then $180°$ out of phase, in a linear summer. The output waveform is simply the algebraic sum of V_1 and V_2. The schematic diagram for the linear summing circuit used in this section is shown in Figure 5-2. Op amps 1 and 2 and potentiometer R_4 set the amplitude and phase of V_1 and V_2. Op amp 3 is a linear inverting summer which simply adds V_1 to V_2 and then inverts their sum, (i.e., $V_{out} = -[V_1 + V_2]$).

Procedure

1. Construct the linear circuit shown in Figure 5-2.
2. Adjust the amplitude of the signal generator output for a 4 Vp–p 1-kHz sine wave at the output of op amp 1 (V_1).

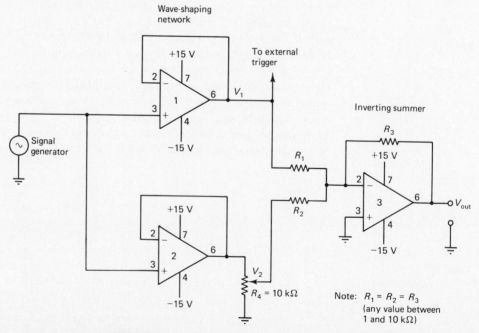

FIGURE 5-2 Linear summing of two in phase sine waves

3. Adjust $R4$ until V_2 is a 2 Vp–p sine wave.

4. Sketch the waveforms for V_1, V_2, and V_{out}.

5. Describe V_{out} in terms of its frequency content, shape, and amplitude.

6. While observing V_{out}, slowly vary R_4 throughout its entire range.

7. Describe what effect varying R_4 has on V_1, V_2, and V_{out}.

8. Change the configuration of op amp 2 to a unity-gain inverting amplifier as shown in Figure 5-3.

9. Repeat steps 2 through 7.

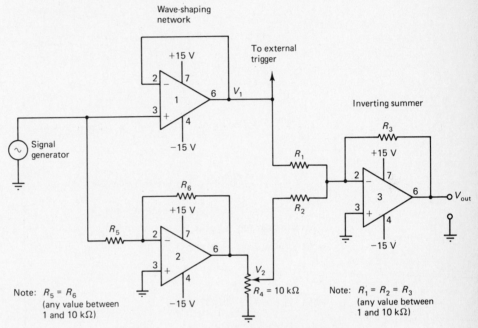

FIGURE 5-3 Linear summing of two out of phase sine waves

SECTION B Linear Summing of a Sine Wave and a Square Wave

In this section a sine wave and a square wave with equal periods (V_1 and V_2) are combined, first in phase and then 180° out of phase, in a linear inverting summer. Again, the output waveform is simply the inverted algebraic sum of V_1 and V_2. The schematic diagram for the linear summing circuit used in this section is shown in Figure 5-4. This circuit is identical to the circuit used in Section A except op amp 2 is a voltage comparator, which converts V_2 to a square wave.

Procedure

1. Construct the linear summer circuit shown in Figure 5-4.

2. Adjust the amplitude of the signal generator output for a 4 Vp–p 1-kHz sine wave at the output of op amp 1 (V_1).

3. Adjust R_4 until V_2 is a 4 Vp–p square wave.

4. Sketch the waveforms for V_1, V_2, and V_{out}.

5. Describe V_{out} in terms of its frequency content, shape, and amplitude.

6. While observing V_{out}, slowly vary R_4 throughout its entire range.

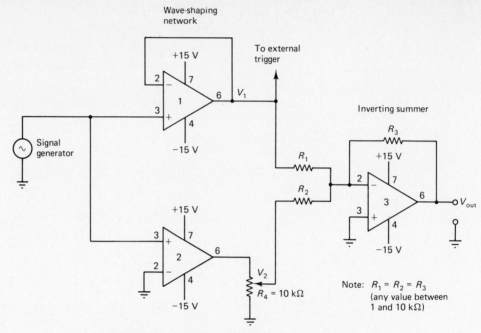

FIGURE 5-4 Linear summing of a sine wave and a square wave

7. Describe what effect varying R_4 has on V_1, V_2, and V_{out}.
8. Reverse the inverting and noninverting inputs to op amp 2.
9. Repeat steps 2 through 7.

SECTION C   Summary

Write a brief summary of the concepts presented in this experiment on linear summing. Include the following items:

1. The concept of algebraic addition of two waveforms in the time domain.
2. The concept of linear mixing two signals.
3. The concept of in- and out-of-phase signals.
4. The concept of time, frequency, and phase coherency.

EXPERIMENT 5   Answer Sheet

NAME: _____ CLASS: _____ DATE: _____

SECTION A

4.

Vertical sensitivity _____ V/cm

Time base _____ sec/cm

Vertical sensitivity _____ V/cm

Time base _____ sec/cm

Vertical sensitivity _____ V/cm

Time base _____ sec/cm

5. _____

_____

_____

7. _____

_____

_____

_____

9.

Vertical sensitivity _____ V/cm

Time base _____ sec/cm

Vertical sensitivity _____ V/cm

Time base _____ sec/cm

Vertical sensitivity _____ V/cm

Time base _____ sec/cm

SECTION B

4.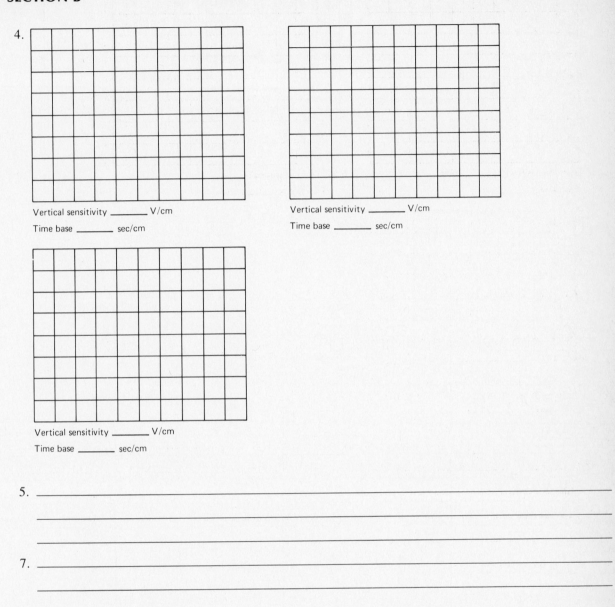

Vertical sensitivity _____ V/cm

Time base _____ sec/cm

Vertical sensitivity _____ V/cm

Time base _____ sec/cm

Vertical sensitivity _____ V/cm

Time base _____ sec/cm

5. _____

_____

_____

7. _____

_____

_____

9.

Vertical sensitivity _____ V/cm

Time base _____ sec/cm

Vertical sensitivity _____ V/cm

Time base _____ sec/cm

Vertical sensitivity _____ V/cm

Time base _____ sec/cm

NONLINEAR AMPLIFICATION AND HARMONIC DISTORTION

REFERENCE TEXT: *Fundamentals of Electronic Communications Systems.*

1. Chapter 1, Harmonic Distortion.
2. Chapter 3, Nonlinear Mixing—Single Input Frequency.

OBJECTIVES

1. To observe the frequency response characteristics of a passive bandstop filter.
2. To observe voltage amplification in a linear amplifier.
3. To observe nonlinear amplification of a single frequency.
4. To observe harmonic distortion caused by nonlinear amplification.

INTRODUCTION

Nonlinear amplification produces harmonic distortion. When a small signal transistor amplifier is overdriven (i.e., operated near saturation or cutoff), the output waveform contains the original input frequency plus its harmonics. If the original input frequency is removed (i.e., filtered out), only the higher harmonics remain, of which the second harmonic is the most significant. In this exercise a class A transistor amplifier is intentionally overdriven. A filter, placed in series with the output of the amplifier, removes a significant portion of the fundamental frequency. Therefore, the final output spectrum contains components of both the fundamental and second harmonic frequencies that are approximately equal in amplitude. The block diagram of the test circuit used in this experiment is shown in Figure 6-1.

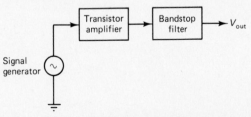

FIGURE 6-1 Nonlinear amplification test circuit

MATERIALS REQUIRED

Equipment:

1 — protoboard
1 — dc power supply (20 V dc)
1 — audio signal generator
1 — standard oscilloscope (10 MHz)
1 — assortment of test leads and hookup wire

Parts List:

1 — small-signal transistor (3904 or equivalent)
3 — 10 k-ohm resistors
1 — 22 k-ohm resistor
1 — 47 k-ohm resistor
2 — 100 k-ohm resistors
2 — 1000-pF capacitors
1 — 2000-pF capacitor
1 — 0.05-μF capacitor
1 — bypass capacitor (1 μF or larger)

SECTION A   Bandstop Filter

In this section the frequency response of a twin-T bandstop filter is examined. A bandstop filter (sometimes called a bandreject or notch-out filter) is the combination of a lowpass and a highpass filter. A bandstop filter blocks all frequencies that fall between the break frequencies of the lowpass and highpass filters. The lower limit of the notch is the upper cutoff frequency of the lowpass filter, and the upper limit is the lower cutoff frequency of the highpass filter. The frequency that falls in the middle of the notch is attenuated the most and is called the center frequency. The schematic diagram for the bandstop filter circuit used in this section is shown in Figure 6-2.

Procedure

1. Construct the twin-T bandstop filter circuit shown in Figure 6-2.
2. Calculate the center frequency of the filter using the following formula.

$$F_c = \frac{1}{2\pi RC}$$

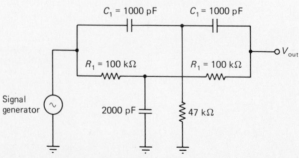

FIGURE 6-2   Twin-T bandstop filter

where F_c = center frequency
$C = C_1$
$R = R_1$

3. Measure the actual center frequency by setting the signal generator output to 4 V p–p and varying its output frequency from 100 Hz to 10 kHz.

4. Plot the frequency response curve for the bandstop filter on semilog graph paper.

5. Calculate the percentage error between the calculated and actual center frequencies using the following formula.

$$\% \text{ error } = \frac{\text{measured value } - \text{ calculated value}}{\text{calculated value}} \times 100$$

6. Calculate the gain of the filter at the center frequency using the following formula.

$$A_v = \frac{V_{\text{out}}}{V_{\text{in}}}$$

where A_v = voltage gain
V_{out} = filter output voltage
V_{in} = filter input voltage

SECTION B Linear Amplification

In this section the voltage gain of a linear class A transistor amplifier is determined by applying an input signal of known amplitude and measuring its amplitude at the output. The amplitude of the input signal must be sufficiently low to ensure that the amplifier is not overdriven (i.e., the transistor is operating over a linear portion of its operating curve). The schematic diagram for the linear transistor amplifier circuit used in this section is shown in Figure 6-3.

Procedure

1. Construct the transistor amplifier circuit shown in Figure 6-3.

2. Set the audio signal generator output to 40 mV p–p and adjust its frequency to the value determined in step 3 of Section A. If the amplitude of the signal generator will not go this low, use a voltage divider to reduce it.

3. Sketch the base input and collector output waveforms.

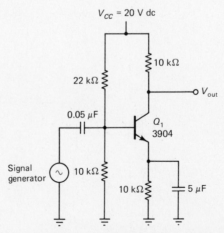

FIGURE 6-3 Linear transistor amplifier

4. Describe the output waveform in terms of frequency content, shape, amplitude, and phase.

5. Measure the gain of the amplifier using the following formula.

$$A_v = \frac{V_{out}}{V_{in}}$$

where A_v = voltage gain
V_{out} = ac collector voltage
V_{in} = ac base voltage

SECTION C   Nonlinear Amplification and Harmonic Distortion

In this section the filter constructed in Section A is placed in series with the output of the amplifier constructed in Section B. An input signal is applied to the amplifier with sufficient amplitude to drive the transistor into a nonlinear portion of its operating curve and produce a distorted signal at the collector. The bandstop filter blocks a significant portion of the fundamental frequency while passing the higher harmonics. Therefore, the waveform at the output of the filter contains the fundamental frequency and its second harmonic with approximately equal amplitudes. The schematic diagram for the test circuit used in this section is shown in Figure 6-4.

Procedure

1. Connect a jumper wire between the collector of Q_1 and the input to the twin-T filter as shown in Figure 6-4.

2. Set the audio signal generator output to 40 mV p–p and adjust its frequency to the center frequency of the filter.

3. Increase the amplitude of the signal generator output voltage until the waveform at the collector of Q_1 distorts significantly.

4. Sketch the waveforms observed at the collector of Q_1 and the output of the filter. The waveform at the output of the filter should be the summation of the input frequency and its second harmonic. You may have to adjust the frequency and amplitude of the signal generator slightly, dep ending on the stability of the generator.

5. Describe the waveforms observed at the collector of Q_1 and the output of the filter in terms of frequency content, shape, amplitude, and phase.

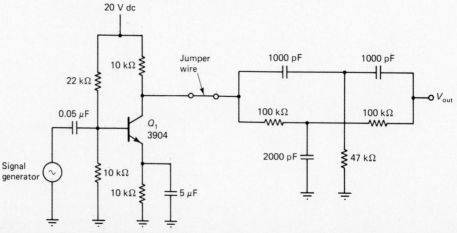

FIGURE 6-4   Nonlinear transistor amplifier-harmonic distortion.

SECTION D Summary

Write a brief summary of the concepts presented in this experiment on nonlinear amplification. Include the following items:

1. The concepts of linear and nonlinear amplification.
2. The relationship between nonlinear amplification and nonlinear distortion.
3. The result of band limiting a signal.
4. The effect that harmonic distortion has on a waveform in both the frequency and time domains.

EXPERIMENT 6   Answer Sheet

NAME: _____ CLASS: _____ DATE: ____

SECTION A

2. F_c (calculated) = _____   3. F_v (measured) = _____

5. % error = _____   6. A_v = _____

SECTION B

3.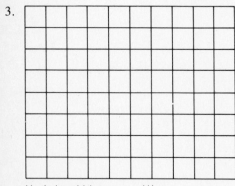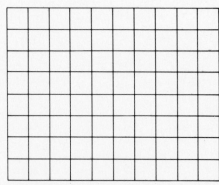

Vertical sensitivity _____ V/cm   Vertical sensitivity _____ V/cm

Time base _____ sec/cm   Time base _____ sec/cm

4. _____

_____

_____

5. A_v = _____

SECTION C

4.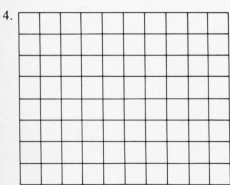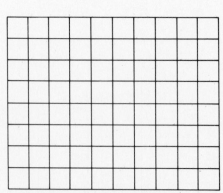

Vertical sensitivity _____ V/cm   Vertical sensitivity _____ V/cm

Time base _____ sec/cm   Time base _____ sec/cm

5. _____

_____

_____

NONLINEAR MIXING

REFERENCE TEXT: *Fundamentals of Electronic Communications Systems.*

1. Chapter 1, Intermodulation Distortion.
2. Chapter 3, Nonlinear Mixing—Multiple Input Frequencies.

OBJECTIVES

1. To observe the frequency response characteristics of a passive bandpass filter.
2. To observe the gain-versus-frequency response characteristics of a tuned transistor amplifier.
3. To observe nonlinear mixing of two sine waves to produce cross product frequencies.

INTRODUCTION

Nonlinear mixing occurs when two or more frequencies are combined in a nonlinear device. Nonlinear mixing produces cross-product frequencies (i.e., sum and difference frequencies). If the cross products are undesired, it is called intermodulation distortion, which is a form of nonlinear distortion. However, in communications circuits it is often necessary or desirable to combine two or more frequencies in a nonlinear device to produce a third frequency or band of frequencies. For example, in modulator circuits, nonlinear mixing is used to up-convert low frequency information signals to higher radio frequencies for transmission, and in receivers, nonlinear mixing is used to down-convert radio frequencies to intermediate frequencies for amplification and band limiting (filtering). In this exercise two relatively high frequency signals are applied to the input of a tuned transistor amplifier. The amplitudes of the input signals are made sufficiently high to drive the amplifier into nonlinear operation, which produces the sum and difference frequencies. The difference frequency is separated from the composite waveform and displayed on a standard oscilloscope. The block diagram of the test circuit used in this experiment is shown in Figure 7-1.

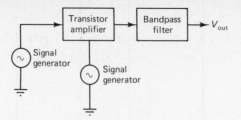

FIGURE 7-1 Nonlinear mixing test circuit

MATERIALS REQUIRED

Equipment:

1 — protoboard
1 — dc power supply (12 V dc)
2 — medium-frequency signal generators (1 MHz)
1 — standard oscilloscope (10 MHz)
1 — assortment of test leads and hookup wire

Parts List:

1 — small-signal transistor (3904 or equivalent)
1 — 3-mH inductor
2 — 1.2 k-ohm resistors
1 — 1 k-ohm resistor
1 — 10 k-ohm resistor
1 — 47 k-ohm resistor
2 — 0.001-μF capacitors
1 — 0.002-μF capacitor
1 — 0.01-μF capacitor

SECTION A   Bandpass Filter

In this section the frequency response of a bandpass filter is examined. A bandpass filter is the combination of a lowpass and a highpass filter. A bandpass filter passes all frequencies that fall between the break frequencies of the lowpass and highpass filters. The lower limit of the passband is equal to the lower cutoff frequency of the highpass filter, and the upper limit of the passband is equal to the upper cutoff frequency of the lowpass filter. The schematic diagram for the bandpass filter circuit used in this section is shown in Figure 7-2.

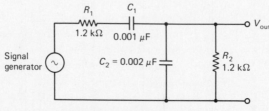

FIGURE 7-2 Bandpass filter

Procedure

1. Construct the bandpass filter circuit shown in Figure 7-2.
2. Calculate the lower break frequency using the following formula.

$$F_L = \frac{1}{2\pi RC}$$

where F_L = lower break frequency
$$R = R_1 + R_2$$
$$C = C_1$$

3. Calculate the upper break frequency using the following formula.

$$F_U = \frac{1}{2\pi RC}$$

where F_U = upper break frequency
$$R = R_{th} = \frac{R_1 \times R_2}{R_1 + R_2}$$
$$C = C_2$$

4. Measure the actual upper and lower break frequencies by setting the signal generator output to 4 V p–p and varying its output frequency from 50 to 150 kHz.
5. Plot the frequency response curve for the bandpass filter on semilog graph paper.
6. Calculate the percentage error between the calculated and actual upper and lower break frequencies using the following formula.

$$\% \text{ error} = \frac{\text{measured value} - \text{calculated value}}{\text{calculated value}} \times 100$$

SECTION B Tuned Amplifier

In this section the resonant frequency of a tuned amplifier is examined. The voltage gain of a tuned amplifier is frequency dependent. That is, certain frequencies are amplified more than others. The tuned amplifier used in this experiment is a transistor amplifier with a tank circuit in the collector. Therefore, the ac collector resistance and voltage gain of the amplifier are maximum at the resonant frequency. A tuned amplifier has a frequency response curve similar to that of a bandpass filter except with gain. The schematic diagram for the tuned amplifier circuit used in this section is shown in Figure 7-3.

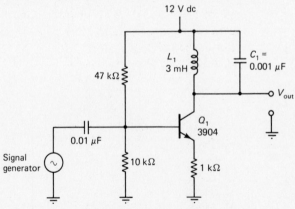

FIGURE 7-3 Tuned amplifier

Procedure

1. Construct the tuned amplifier circuit shown in Figure 7-3.
2. Calculate the resonant frequency of the tank circuit using the following formula.

$$F_r = \frac{1}{2\pi\sqrt{LC}}$$

   where F_r = resonant frequency
   $L = L_1$
   $C = C_1$

3. Measure the actual resonant frequency by setting the signal generator output to 4 V p–p and varying its output frequency from 20 to 100 kHz.
4. Plot the gain-versus-frequency response curve for the tuned amplifier on semilog graph paper.
5. Calculate the percentage error between the calculated and actual resonant frequencies.

SECTION C   Nonlinear Mixing

In this section two signals are applied to a tuned amplifier; one to the base and one to the emitter. The amplitudes of the input signals are sufficiently high to overdrive the amplifier. Therefore, nonlinear mixing occurs and sum and difference frequencies are produced. The amplifier is tuned to the difference frequency, and a bandpass filter in the output circuit removes the two original input frequencies and the sum frequency. The schematic diagram for the test circuit used in this section is shown in Figure 7-4.

Procedure

1. Connect a jumper wire between the collector of Q_1 and the input to the bandpass filter as shown in Figure 7-4.
2. Set the signal generator connected to the base of Q_1 to an output voltage of 13 V p–p and a frequency of 1 MHz.
3. Set the signal generator connected to the emitter of Q_1 to an output voltage of

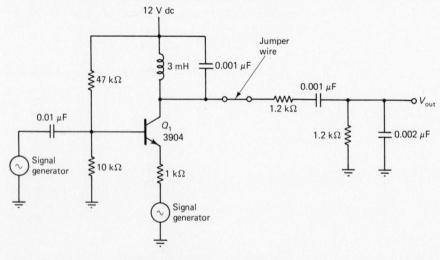

FIGURE 7-4   Nonlinear mixing

7 V p–p and a frequency equal to 1 MHz plus the resonant frequency of the tank circuit.

4. Vary the frequency of the signal generator connected to the emitter of Q_1 until a frequency equal to the difference between the two signal generator frequencies is obtained at the output of the filter.

5. Sketch the waveforms observed at the collector of Q_1 and the output of the filter.

6. Contrast the two waveforms sketched in step 5 in terms of frequency content and amplitude.

SECTION D   Summary

Write a brief description of the concepts presented in this experiment on nonlinear mixing. Include the following items.

1. The frequency response characteristics of a passive bandpass filter.
2. The gain-versus-frequency response characteristics of a tuned amplifier.
3. The effects of combining two signals in a nonlinear amplifier.
4. The process of frequency conversion using nonlinear mixing.

EXPERIMENT 7   Answer Sheet

NAME: _____ CLASS: _____ DATE: _____

SECTION A

2. F_L (calculated) = _____   3. F_U (calculated) = _____

4. F_L (measured) = _____   6. % error (F_L) = _____

    F_U (measured) = _____   % error (F_U) = _____

SECTION B

2. F_r (calculated) = _____   3. F_r (measured) = _____

5. % error = _____

SECTION C

5.

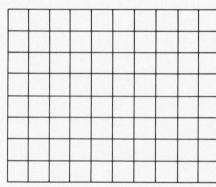

Vertical sensitivity _____ V/cm   Vertical sensitivity _____ V/cm

Time base _____ sec/cm   Time base _____ sec/cm

6. _____

_____

_____

WIEN-BRIDGE OSCILLATOR

REFERENCE TEXT: *Fundamentals of Electronic Communications Systems.*
Chapter 2, Wien-bridge Oscillator.

OBJECTIVES

1. To observe the operation of a feedback oscillator.
2. To observe the effects of regenerative feedback.
3. To observe the operation of a RC phase-shift network.
4. To observe the effects of automatic gain control.
5. To observe the operation of a closed-loop feedback system.

INTRODUCTION

An oscillator is a closed-loop feedback system; a portion of the output signal is fed back to the input. The net gain around the loop must be unity or greater, and the net phase shift must be a positive-integer multiple of 360°. The Wien-bridge oscillator is a free-running RC phase-shift oscillator. It is a relatively stable low-frequency oscillator that is easily tuned over the range of approximately 5 Hz to 1 MHz. In this exercise, a Wien-bridge oscillator is constructed with an operational amplifier, which provides the gain necessary for sustained oscillations. A path for positive (regenerative) feedback is provided by an RC lead-lag network (i.e., the feedback or β network). The lead-lag network also determines the frequency of oscillation by providing a feedback signal with the required amplitude and phase at only one frequency. The block diagram for the oscillator used in this exercise is shown in Figure 8-1.

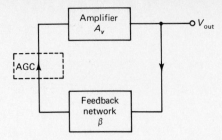

FIGURE 8-1 Oscillator block diagram

MATERIALS REQUIRED

Equipment:

1 — protoboard
1 — dual dc power supply ($+15$ V dc and -15 V dc)
1 — audio signal generator
1 — standard oscilloscope (10 MHz)
1 — assortment of test leads and hookup wire

Parts List:

1 — operational amplifier (741C or equivalent)
1 — low-power diode (1N4004 or equivalent)
1 — JFET (2N5457 or equivalent)
1 — 5 k-ohm variable resistor
1 — 1 k-ohm resistor
2 — 10 k-ohm resistors
1 — 100 k-ohm resistor
1 — 1 M-ohm resistor
2 — 0.01-μF capacitors
2 — 0.1-μF capacitors

SECTION A   Wien-bridge Oscillator

In this section a simple Wien-bridge oscillator without automatic gain control is examined. An op amp noninverting amplifier is used to provide the loop gain, and an RC lead-lag network is used for the feedback network. The schematic diagram for the Wien-bridge oscillator circuit used in this section is shown in Figure 8-2a. R_1, R_2, C_1, and C_2 make up the lead-lag feedback network, which provides 0° phase shift at the frequency of oscillation, F_o (-45° across R_1 and C_1 and $+45$° shift across R_2 and C_2). At F_o, the feedback ratio for the network $\beta = 1/3$. Therefore, F_o is reduced by a factor of three as it passes through the network. This loss is compensated for with the noninverting amplifier, which has a voltage gain $A_v = 3$. Therefore, the total closed-loop gain ($A_v\beta$) at F_o is unity ($3 \times 1/3 = 1$), the net phase shift around the loop is 0°, and oscillations are self-sustained. For all other frequencies, the loop gain is either less than unity and insufficient for self-sustained oscillations, or greater than unity, which causes the circuit to saturate. A portion of V_{out} is fed back through the feedback network with 0° phase shift and then amplified by the noninverting amplifier. Therefore, the total phase shift around the loop is 0°. At very high frequencies, C_1 and C_2 are short circuits, shunting the feedback signal to ground; and

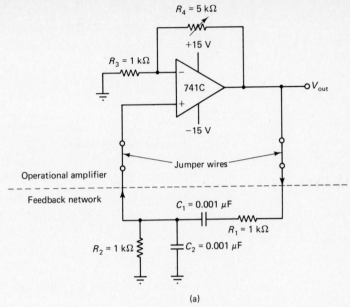

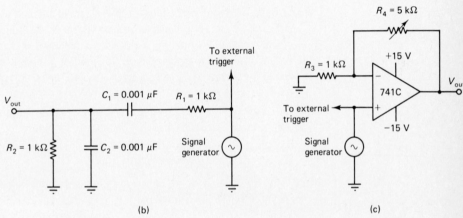

FIGURE 8-2 Wien-bridge oscillator. (a) Schematic diagram. (b) Feedback network phase shift measurements. (c) Op amp voltage gain measurements.

at low frequencies, C_1 and C_2 are open circuits, preventing the feedback signal from reaching the input to the op amp.

Procedure

1. Construct the Wien-bridge oscillator circuit shown in Figure 8-2a (leave the two jumper wires between the feedback network and the op amp disconnected).
2. Calculate F_o for the feedback network using the following formula.

$$F_o = \frac{1}{2\pi RC}$$

where $F_o = 0°$ phase-shift frequency
$R = R_1$
$C = C_1$

3. Connect the signal generator to the input of the feedback network as shown in Figure 8-2b. Adjust the signal generator output amplitude to 4 V p–p, and set its output frequency to the value calculated in step 2.

4. Observe the phase shift across the network with an oscilloscope (on most oscilloscopes, phase-shift measurements require an external trigger).

5. Carefully adjust the signal generator frequency until the input and output signals are in phase.

6. Determine the feedback ratio using the following formula.

$$\beta = \frac{V_{out}}{V_{in}}$$

where β = feedback ratio at F_o
V_{out} = feedback network output voltage
V_{in} = feedback network input voltage

7. Sketch the input and output waveforms for the feedback network.

8. Contrast the feedback network input and output waveforms in terms of frequency, phase, and amplitude.

9. Vary the signal generator frequency while observing the phase shift and feedback ratio.

10. Describe what effect varying the signal generator frequency has on the phase shift and feedback ratio.

11. Use the following procedure to set the gain of the noninverting amplifier to a value equal to $1/\beta$, where β equals the feedback ratio determined in step 6. Disconnect the signal generator from the input to the feedback network and connect it to the noninverting input to the op amp as shown in Figure 8-2c. Set the signal generator output frequency to F_o and adjust its amplitude to 0.5 V p–p. Adjust R_4 for a voltage gain $A_v = 1/\beta$.

12. Disconnect the signal generator and connect the two jumper wires between the op amp and the feedback network as shown in Figure 8-2a.

13. Sketch the waveforms observed at the input and output of the op amp.

14. What is the frequency of oscillation?

15. Vary R_4 throughout its entire range and describe what effect it has on the two waveforms sketched in step 13.

SECTION B   Wien-bridge Oscillator with Automatic Gain

The Wien-bridge oscillator circuit tested in section A will oscillate over a relatively wide range of frequencies. Also, its long-term stability is relatively poor owing to component tolerances and environmental variations. A more constant frequency of oscillation is obtained by using an automatic gain control (AGC) circuit to automatically adjust the gain of the amplifier to compensate for component variations. The Wien-bridge oscillator circuit shown in Figure 8-3 is very similar to the one discussed previously except with the addition of an AGC circuit. The feedback network is identical, consisting of lead-lag network R_1C_1 and R_2C_2. One of the resistors in the amplifier has been replaced with a JFET, which acts like a voltage-controlled resistance. The half-wave rectifier samples the output waveform and produces a negative dc voltage that is proportional to the amplitude of the output signal. The dc voltage provides a negative gate-source bias, which controls the source-to-drain resistance of the JFET. When there is no output voltage, the bias is 0, the resistance of the JFET is low, and the gain of the amplifier is high. When the output voltage increases, the bias voltage goes more negative, which increases the JFET resistance and reduces the gain of the amplifier. Thus, the closed-loop gain is automatically adjusted to compensate for loop variations.

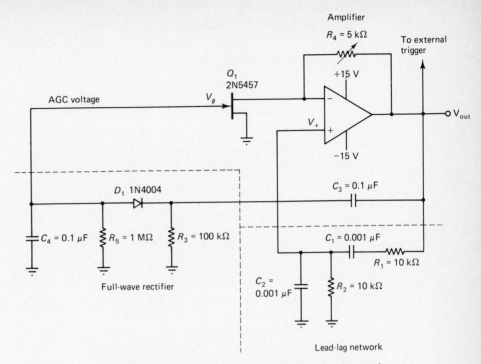

FIGURE 8-3 Wien-bridge oscillator with automatic gain control.

Procedure

1. Construct the Wien-bridge oscillator circuit shown in Figure 8-3.
2. Calculate the frequency of oscillation using the following formula.

$$F_o = \frac{1}{2\pi RC}$$

where F_o = frequency of oscillation
$C = C_1$
$R = R_1$

3. Observe the output waveform and adjust R_4 for an output signal with minimum distortion.
4. What is the frequency of oscillation?
5. Calculate the percentage error between the calculated and actual frequencies of oscillation using the following formula.

$$\% \text{ error} = \frac{\text{measured value } - \text{ calculated value}}{\text{calculated value}} \times 100$$

6. Sketch the waveforms at V_{out}, V_+, and V_g.
7. Describe the waveforms sketched in step 6 in terms of amplitude, phase, and circuit gain.
8. Vary R_4 throughout its entire range while observing V_{out}, V_+, and V_g.
9. Describe what effect varying R_4 has on the frequency, amplitude, and phase of V_{out}, V_+, and V_g.
10. Design a Wien-bridge oscillator with a frequency of oscillation $F_o = 4.8$ kHz. Use capacitance values $C_1 = C_2 = 0.01$ μF.
11. Construct the oscillator circuit designed in step 10.

SECTION C   Summary

Write a brief summary of the concepts presented in this experiment on Wien-bridge oscillators. Include the following items:

1. The basic operation of an RC feedback oscillator.
2. The relationship between the amplifier gain and the feedback ratio.
3. The phase relationship between the output and feedback signals.
4. The gain and phase characteristics of the feedback network.
5. The basic operation of the automatic gain mechanism used in section B.

EXPERIMENT 8   Answer Sheet

NAME: _____ CLASS: _____ DATE: _____

SECTION A

2. F_0 (calculated) = _____   6. β = _____

7.

Vertical sensitivity _____ V/cm
Time base _____ sec/cm

Vertical sensitivity _____ V/cm
Time base _____ sec/cm

8. _____

_____

_____

10. _____

_____

_____

13.

Vertical sensitivity _____ V/cm
Time base _____ sec/cm

Vertical sensitivity _____ V/cm
Time base _____ sec/cm

14. F_o (measured) = _____

15. _____

_____

_____

SECTION B

2. F_o (calculated) = _____

4. F_o (measured) = _____

5. % error = _____

6.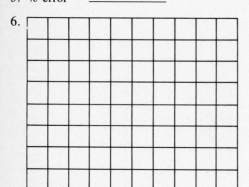

Vertical sensitivity _____ V/cm

Time base _____ sec/cm

Vertical sensitivity _____ V/cm

Time base _____ sec/cm

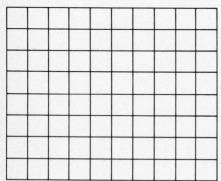

Vertical sensitivity _____ V/cm

Time base _____ sec/cm

7. _____

_____

_____

9. _____

_____

_____

10. $R_1 = R_2 =$ _____

COLPITTS OSCILLATOR

REFERENCE TEXT: *Fundamentals of Electronic Communications Systems.*

1. Chapter 2, LC Oscillators—Oscillator Action.
2. Chapter 2, Colpitts Oscillator.

OBJECTIVES

1. To observe the gain-versus-frequency response characteristics of a tuned class A transistor amplifier.
2. To observe the operation of an LC feedback oscillator.
3. To observe the operation of a Colpitts oscillator.

INTRODUCTION

An oscillator is a circuit that produces a repetitive output waveform. The waveform can be a sine wave, a square wave, or any other type of wave so long as it is repetitive. If the oscillator is self-sustaining, an external input signal is not required. In essence, a feedback oscillator is an amplifier in which a portion of the output signal is fed back to the input with the proper amplitude and phase to sustain oscillations. According to the Barkhausen criterion, for a feedback circuit to sustain oscillations, the net gain around the feedback loop must be unity or greater and the net phase shift around the loop must be a positive-integer multiple of 360° (i.e., 0°). There are four basic requirements for a feedback oscillator to work: amplification, positive (regenerative) feedback, frequency dependency, and a power source. In this exercise a Colpitts oscillator is constructed using a tuned class A transistor amplifier. The amplifier provides the necessary loop gain, and a tank circuit in the collector determines the frequency of oscillation. The feedback signal is taken from a capacitive voltage divider in the tank circuit. The block diagram for the test circuit used in this experiment is shown in Figure 9-1.

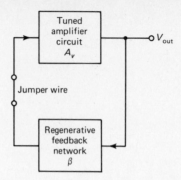

FIGURE 9-1 Colpitts oscillator block diagram.

MATERIALS REQUIRED

Equipment:

1 — protoboard

1 — dc power supply (12 V dc)

1 — medium-frequency signal generator (1.5 MHz)

1 — standard oscilloscope (10 MHz)

1 — assortment of test leads and hookup wire

Parts List:

1 — small signal transistor (3904 or equivalent)

1 — radio frequency choke (RFC) 1 to 3 mH

1 — 7.1- to 12.5-μH variable inductor

1 — 22-ohm resistor

1 — 470-ohm resistor

1 — 1 k-ohm resistor

1 — 2.7 k-ohm resistor

1 — 8.2 k-ohm resistor

1 — 0.0022-μF capacitor

1 — 0.022-μF capacitor

1 — 0.047-μF capacitor

1 — 0.1-μF capacitor

SECTION A   Tuned Amplifier

In this section the gain-versus-frequency response characteristics of a tuned class A amplifier are examined. The voltage gain of a tuned amplifier is frequency dependent; a narrow band of frequencies is amplified more than other frequencies. The frequency response of a tuned amplifier is similar to that of a bandpass filter except with gain. With a tuned amplifier, maximum voltage gain is achieved at the resonant frequency of the tank circuit. The schematic diagram for the class A tuned amplifier circuit used in this section is shown in Figure 9-2. The resonant frequency of the tank circuit is tunable with variable inductor L_1.

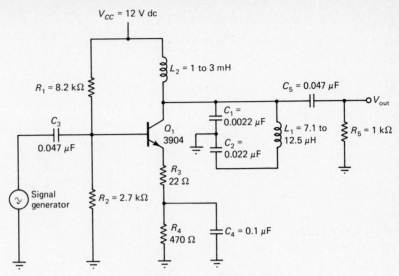

FIGURE 9-2 Tuned amplifier

Procedure

1. Construct the class A tuned amplifier circuit shown in Figure 9-2.
2. With L_1 set to its maximum position, calculate the resonant frequency of the tank circuit using the following formula.

$$F_r = \frac{1}{2\pi\sqrt{LC}}$$

where F_r = resonant frequency
$L = L_1$
$$C = \frac{C_1 \times C_2}{C_1 + C_2}$$

3. Measure the actual resonant frequency by setting the signal generator output amplitude to 200 mV p–p and varying its output frequency from 500 kHz to 1.5 MHz.
4. Calculate the percentage error between the calculated and actual resonant frequencies.
5. Measure the voltage gain of the tuned amplifier at resonance.
6. Repeat steps 2 through 5 with L_1 set to both its midrange and minimum positions.

SECTION B Colpitts Feedback Oscillator

In this section the operation of a Colpitts feedback oscillator is examined. The schematic diagram for the Colpitts oscillator circuit used in this section is shown in Figure 9-3. This circuit is identical to the circuit shown in Figure 9-2 except for the addition of jumper wire J_1, connected between the top of the tank circuit and coupling capacitor C_3. Q_1 and its associated circuitry is a class A voltage amplifier which provides the voltage gain necessary for self-sustained oscillations to occur. The frequency of oscillation is determined by the resonant frequency of the output tank circuit (L_1, C_1, and C_2). A portion of the output signal is taped off capacitor C_2 and fed back to the input of the voltage amplifier through coupling capacitor C_3. The overall closed-loop gain must be unity at the tank circuit reso-

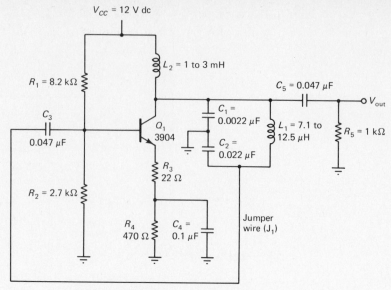

FIGURE 9-3 Colpitts oscillator

nant frequency. That is, the product of the amplifier gain and the loss through the feed-back network must be unity.

Procedure

1. Construct the Colpitts feedback oscillator circuit shown in Figure 9-3 (set inductor L_1 to midrange).

2. Calculate the feedback ratio using the following formula.

$$\beta = \frac{X_{C_2}}{X_{C_1}} = \frac{C_1}{C_2}$$

where β = feedback ratio

3. Calculate the amplifier voltage gain necessary for an overall closed-loop gain of unity using the following formula (a closed-loop gain of unity is required for self-sustained oscillations to occur).

$$A_v = \frac{1}{\beta} = \frac{X_{C_1}}{X_{C_2}} = \frac{C_2}{C_1}$$

where A_v = voltage gain

4. Measure the ac collector and base voltages for Q_1 and determine the voltage gain from the following formula.

$$A_v = \frac{v_c}{v_b}$$

where A_v = voltage gain
 v_c = ac collector voltage
 v_b = ac base voltage

5. Determine the actual feedback ratio using the following formula.

$$\beta = \frac{v_b}{v_c}$$

where β = feedback ratio
 v_b = ac base voltage
 v_c = ac collector voltage

6. Sketch the waveforms at both the collector and base of Q_1.

7. Calculate the overall closed-loop gain using the following formula.

$$A_{cl} = A_v \beta$$

where A_{cl} = closed-loop gain
A_v = amplifier voltage gain
β = feedback ratio

8. Vary L_1 throughout its entire range and measure the minimum and maximum frequencies of oscillation.

9. Design a Colpitts oscillator circuit with a frequency of oscillations $F_o = 1.6$ MHz. Use an inductance $L = 10$ μH and a feedback ratio $C_1/C_2 = 10{:}1$.

10. Construct the oscillator circuit designed in step 9 and verify its proper operation.

SECTION C   Summary

Write a brief summary of the concepts presented in this experiment on the Colpitts oscillator. Include the following concepts:

1. The basic operation of an LC feedback oscillator.

2. The gain-versus-frequency response characteristics of a tuned class A amplifier.

3. The frequency-determining components of the Colpitts oscillator.

4. The relationship between the transistor amplifier gain and the feedback ratio.

5. The relationship between the phase of the output and feedback signals.

6. The basic operation of the Colpitts oscillator.

EXPERIMENT 9   Answer Sheet

NAME: _____ CLASS: _____ DATE: _____

SECTION A

2. F_r (calculated) = _____

3. F_r (measured) = _____

4. % error = _____

5. A_v = _____

6. F_r (calculated) = _____

   F_r (measured) = _____

   % error = _____

   A_v = _____

SECTION B

2. β (calculated) = _____

3. A_v (calculated) = _____

4. A_v (measured) = _____

5. β (measured) = _____

6.

Vertical sensitivity _____ V/cm

Time base _____ sec/cm

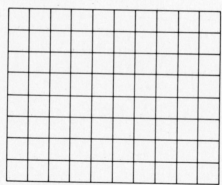

Vertical sensitivity _____ V/cm

Time base _____ sec/cm

7. A_{cl} = _____

9. C_1 = _____

   C_2 = _____

LINEAR INTEGRATED-CIRCUIT FUNCTION GENERATOR VOLTAGE-CONTROLLED OSCILLATOR

REFERENCE TEXT: *Fundamentals of Electronic Communications Systems.*

1. Chapter 2, Large Scale Integration Oscillators.
2. Chapter 5, Voltage Controlled Oscillator.

OBJECTIVES

1. To observe the operation of a voltage-controlled oscillator.
2. To observe the frequency-versus-timing capacitance characteristics of a voltage-controlled oscillator.
3. To observe the frequency-versus-timing resistance characteristics of a voltage-controlled oscillator.
4. To observe the frequency-versus-control voltage characteristics of a voltage-controlled oscillator.
5. To observe the operation of a linear integrated-circuit function generator.

INTRODUCTION

A voltage-controlled oscillator (VCO) is a free-running oscillator with a stable frequency of oscillation that is dependent on a timing capacitance, a timing resistance, and an external control voltage. The output from a VCO is a frequency and its input is a bias or control signal that can be either a dc or an ac voltage. In this experiment the operation of an XR-2206 linear integrated circuit (LIC) function generator is examined. The XR-2206 function generator is a precision monolithic voltage-controlled oscillator that can provide simultaneous sine and square wave outputs or simultaneous triangular and square wave outputs over a frequency range of 0.01 Hz to 1 MHz. Linear integrated-circuit function generators feature excellent temperature stability, low sine wave distortion, linear frequency-versus-amplitude output characteristics, and a wide linear sweep range. Typical applications for LIC function generators are waveform generation, sweep generation, AM and FM generation, voltage-to-frequency conversion, frequency shift keying, and phase-locked loops. The block diagram for the XR-2206 function generator is shown in Figure 10-1. The XR-2206 comprises four functional blocks: a voltage-controlled oscillator, an

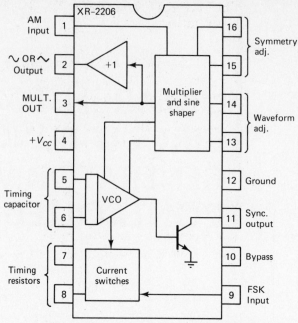

FIGURE 10-1   XR-2206 Functional block diagram

analog multiplier and sine shaper, a unity gain buffer amplifier, and a set of current switches.

MATERIALS REQUIRED

Equipment:

1 — protoboard
1 — dual dc power supply (+ 12 V dc and − 5 V dc to +5 V dc)
1 — audio signal generator
1 — standard oscilloscope (10 MHz)
1 — assortment test leads and hookup wire

Note: If the audio signal generator is equipped with a variable dc offset, the − 5 to +5 V dc supply is not needed.

Parts List:

1 — XR-2206 monolithic function generator
3 — 4.7 k-ohm resistors
1 — 6.8 k-ohm resistor
2 — 10 k-ohm resistors
1 — 22 k-ohm resistor
2 — 47 k-ohm resistors
1 — 100 k-ohm resistor
1 — 1 k-ohm variable resistor
1 — 10 k-ohm variable resistor
1 — 0.001-μF capacitor
1 — 0.01-μF capacitor
1 — 0.1-μF capacitor
2 — 1-μF capacitors
1 — 10-μF capacitor

SECTION A   Triangular Wave, Sine Wave, and Square Wave Generation

In this section the three output functions available with the XR-2206 function generator are examined. The schematic diagram for the function generator used in this section is shown in Figure 10-2. The function generator circuit shown in Figure 10-2 can be used to simultaneously produce either triangular- and square-wave outputs or sine- and square-wave outputs. The free-running oscillator frequency (F_o) is determined by the external timing capacitor (C_1) connected between pins 5 and 6, and by the external timing resistor (R_1) connected between either pin 7 or pin 8 and ground. The frequency is varied by changing the resistance of R_1 or the capacitance of C_1.

Procedure

1. Construct the function generator circuit shown in Figure 10-2.
2. Calculate the free-running oscillator frequency using the following formula.

$$F_o = \frac{1}{RC}$$

    where F_o = free-running frequency
        $R = R_1$
        $C = C_1$

3. Sketch the waveform observed on pin 2 of the function generator.
4. Measure the frequency of the waveform sketched in step 3.
5. Calculate the percentage error between the calculated and actual free-running frequencies using the following formula.

$$\% \text{ error} = \frac{\text{measured value} - \text{calculated value}}{\text{calculated value}} \times 100$$

6. Sketch the waveform observed on pin 11 of the function generator.
7. Measure the frequency of the waveform sketched in step 6.
8. Replace R_2 with a 1 k-ohm variable resistor (rheostat).

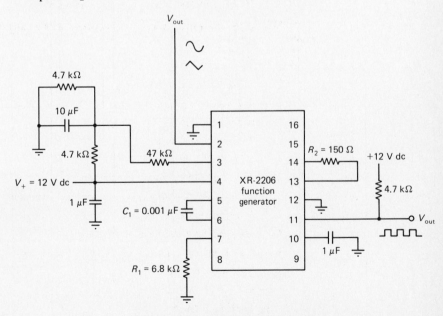

FIGURE 10-2   XR-2206 Function generator

9. Vary R_2 until a sine wave with minimum distortion is observed on pin 2 of the function generator.
10. Sketch the waveform observed in step 9.
11. Measure the frequency of the waveform sketched in step 10.
12. Move R_1 from pin 7 to pin 8 while observing the waveform on pin 2 of the function generator. Describe what effect moving R_1 from pin 7 to pin 8 had on the output waveform.

SECTION B   Output Frequency-versus-Timing Capacitance

In this section the output frequency-versus-timing capacitance characteristics of the XR-2206 function generator are examined. The schematic diagram for the function generator circuit used in this section is shown in Figure 10-3. The VCO produces an output frequency which is proportional to the timing capacitance C_1 between pins 5 and 6 and the input current to either pin 7 or pin 8. The input current is produced by an internal bias voltage and a timing resistor R_1 placed between either pin 7 or pin 8 and ground. If R_1 is held constant, the function generator output frequency is proportional to the capacitance of timing capacitor C_1.

Procedure

1. Construct the oscillator circuit shown in Figure 10-3.
2. Calculate the free-running oscillator frequency.
3. Adjust R_2 until a sine wave with minimum distortion is observed at V_{out}.
4. Measure the frequency of the waveform observed in step 3.
5. Measure the amplitude of the waveform observed in step 3.
6. Repeat steps 2 through 5 using the following values for C_1: 0.01 μF, 0.1 μF, 1 μF, and 10 μF.
7. Construct a graph of the output frequency-versus-timing capacitance for the capacitance values given and frequencies measured in steps 2 through 6.

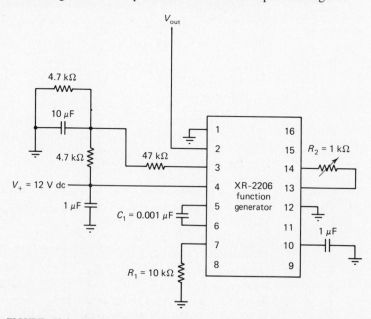

FIGURE 10-3   Output frqeuncy-versus-timing capacitance and output frequency-versus-timing resistance

8. Construct a graph of the output amplitude-versus-frequency characteristics for the frequencies and amplitudes measured in steps 2 through 6.

9. Describe the relationship between the timing capacitance value, the output frequency, and the output amplitude.

SECTION C   Output Frequency-versus-Timing Resistance

In this section the output frequency-versus-timing resistance characteristics of the XR-2206 function generator are examined. The function generator circuit used in this section is identical to the circuit used in Section B and is shown in Figure 10-3. If the capacitance of timing capacitor C_1 is held constant, the function generator output frequency is proportional to the input current to either pin 7 or pin 8, which is a function of the resistance of timing resistor R_1. Consequently, for a fixed capacitance C_1, the function generator output frequency is proportional to the resistance of R_1. In essence, a VCO performs voltage-to-frequency conversion.

Procedure

1. Construct the function generator circuit shown in Figure 10-3.
2. Calculate the free-running oscillator frequency.
3. Adjust R_2 until a sine-wave output with minimum distortion is observed at V_{out}.
4. Measure the frequency of the waveform observed in step 3.
5. Measure the amplitude of the waveform observed in step 3.
6. Repeat steps 2 through 4 using the following values for R_1: 10 k, 22 k, 47 k, and 100 k ohms.
7. Construct a graph of the output frequency-versus-timing resistance for the resistance values given and frequencies measured in steps 2 through 6.
8. Construct a graph of the output amplitude-versus-frequency characteristics for the frequencies and amplitudes measured in steps 2 through 6.
9. Place the 10 k-ohm variable resistor (rheostat) in parallel with R_1.
10. Vary the rheostat's resistance and describe what effect varying it has on the output waveform.
11. Describe the relationship between the timing resistance value, the output frequency, and the output amplitude.

SECTION D   Sweep-Frequency Operation

In this section the operation of the XR-2206 function generator as a sweep-frequency oscillator is examined. The schematic diagram for the sweep-frequency oscillator circuit used in this section is shown in Figure 10-4. As previously stated, the frequency of oscillation of the XR-2206 function generator is proportional to the total timing current (I_T) drawn from either pin 7 or pin 8. Pins 7 and 8 are low-impedance points that are internally biased at $+3$ V with respect to pin 12 (ground). The output frequency varies linearly with timing current over a range of current values from 1 μA to 3 mA. The output frequency can be controlled by applying a control voltage (V_C) to the selected timing pin as shown in Figure 10-4.

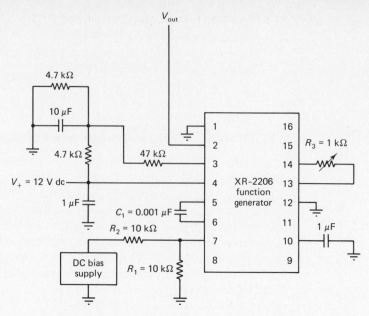

FIGURE 10-4   XR-2206 Function generator sweep frequency operation

Procedure

1. Construct the sweep-frequency oscillator shown in Figure 10-4.
2. Adjust the dc bias supply voltage V_C to 0 V.
3. Adjust R_3 until a sine wave with minimum distortion is observed at V_{out}.
4. Measure the frequency of oscillation and output amplitude for values of V_C from −5 V dc to +5 V dc.
5. Construct a graph of the output frequency-versus-control voltage for the frequencies measured in step 4.
6. Construct a graph of the output amplitude-versus-frequency characteristics for the control voltages given and frequencies measured in step 4.
7. Describe the relationship between the dc bias voltage, the output frequency, and the output amplitude.

SECTION E   Summary

Write a brief summary of the concepts presented in this experiment on linear integrated circuit function generators. Include the following items:

1. The relationship between output frequency and timing capacitance.
2. The relationship between output frequency and timing resistance.
3. The relationship between input currrent and frequency of oscillation.
4. The relationship between control voltage and frequency of operation.
5. The concept of sweep-frequency generation.

EXPERIMENT 10   Answer Sheet

NAME: _____ CLASS: _____ DATE: _____

SECTION A

2. F_o (calculated) = _____

3.

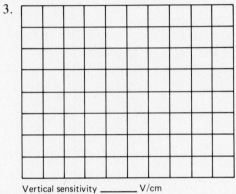

Vertical sensitivity _____ V/cm

Time base _____ sec/cm

4. F_o (measured) = _____   5. % error = _____

6.

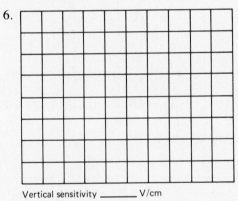

Vertical sensitivity _____ V/cm

Time base _____ sec/cm

7. F_o (measured) = _____

10.

Vertical sensitivity _____ V/cm

Time base _____ sec/cm

11. F_o (measured) = _____

12. _____

_____

_____

SECTION B

2. F_o (calculated) = _____   4. F_o (measured) = _____

5. V_o (measured) = _____

6.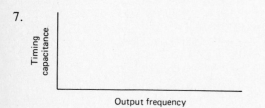

C_1 (μF)	F_o (calculated)	F_o (measured)	V_o (measured)
0.001			
0.01			
0.1			
1			
10			

7.

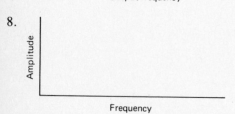

Timing capacitance vs. Output frequency

8.

Amplitude vs. Frequency

9. _____

_____

_____

SECTION C

2. F_o (calculated) = _____   4. F_o (measured) = _____

5. V_o (measured) = _____

6.

R_1 (kΩ)	F_o (calculated)	F_o (measured)	V_o (measured)
10			
22			
47			
100			

7.

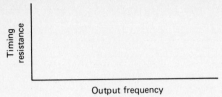

8.

10. _____
    _____
    _____

11. _____
    _____
    _____

SECTION D

4.

V_C (V)	F_o (measured)	V_o (measured)
−5		
−4		
−3		
−2		
−1		
0		
+1		
+2		
+3		
+4		
+5		

5.

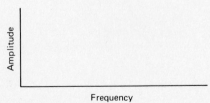

6.

7. _____

_____

_____

FREQUENCY MULTIPLICATION

REFERENCE TEXT: *Fundamentals of Electronic Communications Systems.*
Chapter 2, Frequency Multipliers.

OBJECTIVES

1. To observe the process of frequency multiplication.
2. To observe harmonic generation.

INTRODUCTION

In electronic communications systems it is often necessary to generate harmonic frequencies (i.e., multiples of a base frequency). Synchronous frequency generation can be accomplished with frequency multipliers. A frequency multiplier takes advantage of the nonlinear characteristics of a circuit. When a pure sine wave is distorted, a repetitive waveform is produced, which is made up of a series of harmonically related sine waves with a fundamental frequency equal to the original input frequency. A tuned-circuit frequency multiplier comprises a frequency source (i.e., an oscillator), a nonlinear amplifier, and a frequency selective circuit, such as an LC tank circuit. The base frequency is amplified by the nonlinear amplifier, and the tank circuit separates the desired harmonic from the complex output waveform. The block diagram for the test circuit used in this experiment is shown in Figure 11-1.

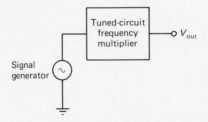

FIGURE 11-1 Frequency multiplier test circuit

MATERIALS REQUIRED

Equipment:

1 — protoboard
1 — dc power supply (15 V dc)
1 — medium-frequency signal generator (1 MHz)
1 — standard oscilloscope (10 MHz)
1 — assortment of test leads and hookup wire

Parts List:

1 — small signal transistor (3904 or equivalent)
1 — 3-mH inductor
1 — 1 k-ohm resistor
1 — 10 k-ohm resistor
1 — 47 k-ohm resistor
2 — 100 k-ohm resistors
1 — 0.001-µF capacitor
2 — 1-µF capacitors

SECTION A   Tuned Amplifier

In this exercise the gain-versus-frequency response characteristics of a class A tuned amplifier are examined. The schematic diagram for the class A tuned amplifier circuit used in this section is shown in Figure 11-2. The maximum voltage gain for the tuned amplifier occurs at the resonant frequency of the tank circuit (L_1 and C_1).

Procedure

1. Construct the class A tuned amplifier circuit shown in Figure 11-2.
2. Calculate the resonant frequency of the tank circuit using the following formula.

$$F_r = \frac{1}{2\pi\sqrt{LC}}$$

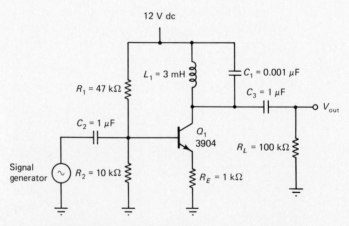

FIGURE 11-2   Frequency multiplier schematic diagram

where F_r = resonant frequency
$L = L_1$
$C = C_1$

3. Measure the actual resonant frequency by setting the signal generator output voltage to 0.5 V p–p and varying its output frequency from 20 to 200 kHz.

4. Calculate the percentage error between the calculated and actual resonant frequencies using the following formula.

$$\% \text{ error} = \frac{\text{measured value} - \text{calculated value}}{\text{calculated value}} \times 100$$

5. Plot the gain-versus-frequency response curve for the tuned amplifier on semilog graph paper.

6. Measure the voltage gain of the amplifier at the resonant frequency using the following formula.

$$A_v = \frac{V_{out}}{V_{in}}$$

where A = voltage gain
V_{out} = ac output voltage
V_{in} = ac base voltage

SECTION B   Frequency Multiplier

In this section the operation of a tuned-circuit frequency multiplier is examined. The frequency multiplier examined is a tuned class A amplifier where the resonant frequency of the tuned circuit is equal to an integer multiple of the input frequency. Because a frequency translation is performed, the voltage gain is called conversion gain. The conversion gain is simply the ratio of the output voltage to the input voltage where the input and output frequencies are not the same. The tune amplifier circuit used in this section is the same circuit that was used in section A and shown in Figure 11-2.

Procedure

1. Construct the class A tuned amplifier circuit shown in Figure 11-2.

2. Set the signal generator output frequency to the value determined in step 3 of Section A and adjust its output voltage to 15 V p–p.

3. Sketch the input and output waveforms.

4. Describe the waveform observed at the output of the tuned amplifier in terms of shape, amplitude, and frequency content.

5. While observing the waveform at the output of the tuned amplifier, reduce the frequency of the signal generator until an undistorted waveform is observed that is equal to twice the signal generator output frequency (the output voltage from the signal generator may have to be readjusted slightly).

6. Sketch the input and output waveforms.

7. Describe the waveform observed at the output of the tuned amplifier in terms of shape, amplitude, and frequency content.

8. Measure the conversion gain of the frequency multiplier using the following formula.

$$A_c = \frac{V_{out}}{V_{in}}$$

where A_v = conversion gain
V_{out} = ac output voltage
V_{in} = ac base voltage

9. While observing the waveform at the output of the tuned amplifier, reduce the frequency of the signal generator until an undistorted waveform is observed that is equal to three times the signal generator output frequency (again, the output voltage from the signal generator may have to be readjusted slightly).

10. Sketch the input and output waveforms.

11. Describe the waveform observed at the output of the tuned amplifier in terms of shape, amplitude, and frequency content.

12. Measure the conversion gain of the frequency multiplier.

13. While observing the waveform at the output of the tuned amplifier, reduce the frequency of the signal generator until an undistorted waveform is observed that is equal to four times the signal generator output frequency.

14. Sketch the input and output waveforms.

15. Describe the waveform observed at the output of the tuned amplifier in terms of shape, amplitude, and frequency content.

SECTION C   Summary

Write a brief summary of the concepts presented in this experiment on frequency multiplication. Include the following items:

1. The operation of a class A tuned circuit.

2. The operation of a tuned-circuit frequency multiplier.

3. The concept of nonlinear amplification.

4. The harmonic composition of a complex repetitive wave.

5. The relationship between the amplitude and frequency of harmonics.

EXPERIMENT 11   Answer Sheet

NAME: _____   CLASS: _____   DATE: _____

SECTION A

2. F_r (calculated) = _____   3. F_r (measured) = _____

4. % error = _____   6. A_v = _____

SECTION B

3.

Vertical sensitivity _____ V/cm

Time base _____ sec/cm

Vertical sensitivity _____ V/cm

Time base _____ sec/cm

4. _____

_____

_____

6.

Vertical sensitivity _____ V/cm

Time base _____ sec/cm

Vertical sensitivity _____ V/cm

Time base _____ sec/cm

7. _____

_____

_____

8. A_c (measured) = _____

10.
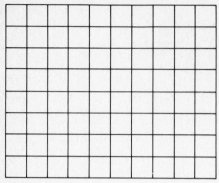
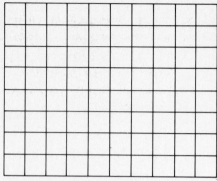

Vertical sensitivity _____ V/cm

Time base _____ sec/cm

Vertical sensitivity _____ V/cm

Time base _____ sec/cm

11. _____

_____

_____

12. A_c (measured) = _____

14.
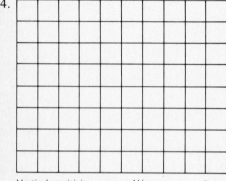
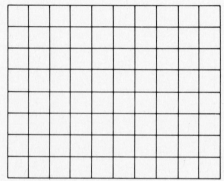

Vertical sensitivity _____ V/cm

Time base _____ sec/cm

Vertical sensitivity _____ V/cm

Time base _____ sec/cm

15. _____

_____

_____

CLASS A, AM DSBFC MODULATOR

REFERENCE TEXT: *Fundamentals of Electronic Communications Systems.*

1. Chapter 3, Low Power AM DSBFC Transistor Modulator.
2. Chapter 3, Trapezoidal Patterns.

OBJECTIVES

1. To observe the characteristics of a class A transistor amplifier.
2. To observe the operation of a simple AM DSBFC modulator using emitter modulation.
3. To observe the operation of a class A transistor modulator.
4. To observe an AM DSBFC envelope and the relationship between the modulated envelope, the modulating signal, and the carrier.
5. To observe trapezoidal patterns for AM signals.

INTRODUCTION

When a relatively high frequency carrier signal is mixed in a nonlinear amplifier with a relatively low frequency modulating signal, amplitude modulation occurs. In this exercise amplitude modulation is accomplished using a class A transistor modulator. A low amplitude carrier signal is applied to the base of the transistor. With no modulating signal input, the amplifier is linear and the output is simply the carrier input signal amplified and inverted. When a modulating signal is applied to the emitter, the Q-point of the transistor is driven toward saturation and cutoff (i.e., into nonlinear operation), causing a proportional change in the dynamic ac emitter resistance (r'_e) and a corresponding change in the voltage gain. In essence, the audio signal modulates (varies or changes) the voltage gain of the amplifier, producing an AM envelope at the output. Because the modulating signal is applied to the emitter, this type of modulation is called emitter modulation. Emitter modulation has several inherent disadvantages, including low power efficiency and low output power. The block diagram for the test circuit used in this experiment is shown in Figure 12-1.

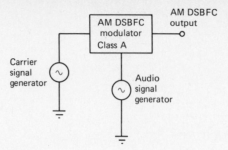

FIGURE 12-1   AM DSBFC modulator test circuit

MATERIALS REQUIRED

Equipment:

1 — protoboard

1 — dc power supply (25 V dc)

1 — audio signal generator

1 — medium-frequency signal generator (1 MHz)

1 — standard oscilloscope (10 MHz)

1 — assortment of test leads and hookup wire

Parts List:

1 — small signal transistor (3904 or equivalent)

1 — 1 k-ohm resistor

1 — 2.2 k-ohm resistor

3 — 10 k-ohm resistors

1 — 22 k-ohm resistor

1 — 0.001-μF capacitor

1 — 0.01-μF capacitor

1 — 0.022-μF capacitor

SECTION A   Class A Transistor Amplifier

In this section the characteristics of a linear class A transistor amplifier are examined. With linear amplification, the output signal is an exact replica of the input signal except amplified and inverted. The schematic diagram for the linear amplifier circuit used in this section is shown in Figure 12-2. This circuit is a Thevenin biased class A common emitter amplifier.

Procedure

1. Construct the common-emitter amplifier circuit shown in Figure 12-2.
2. Calculate the voltage gain of the amplifier using the following formula (assume β = 100).

$$A_v = \frac{r_c}{r'_e}$$

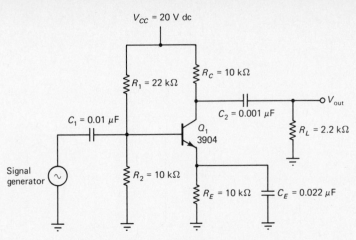

FIGURE 12-2 Class A common emitter amplifier

where A_v = voltage gain
r_c = ac collector resistance
r'_e = dynamic ac emitter resistance

$$r'_e = \frac{25 \text{ mV}}{I_E}$$

and

$$I_E = \frac{V_{th} - V_{be}}{\dfrac{R_{th}}{\beta} + R_E}$$

and

$$r_c = \frac{R_C \times R_L}{R_C + R_L}$$

3. Measure the actual voltage gain of the amplifier by setting the signal generator output to a 100 mVp–p sine wave with a frequency $F = 1$ MHz.
4. Calculate the percentage error between the calculated and actual voltage gains using the following formula.

$$\% \text{ error} = \frac{\text{measured value} - \text{calculated value}}{\text{calculated value}} \times 100$$

5. Sketch the waveforms observed at the base and collector of Q_1.
6. Describe the waveforms sketched in step 5 in terms of amplitude, frequency content, and phase.

SECTION B   Class A, AM DSBFC Modulator

In this section the characteristics of a class A, AM DSBFC transistor modulator are examined. The schematic diagram for the modulator circuit used in this section is shown in Figure 12-3. A relatively high frequency carrier signal is applied to the base of Q_1, and a relatively low frequency modulating signal is applied to the emitter. Therefore, this circuit is an emitter modulator. With emitter modulation, the audio signal voltage adds to and subtracts from the quiescent base-emitter bias voltage, producing corresponding changes

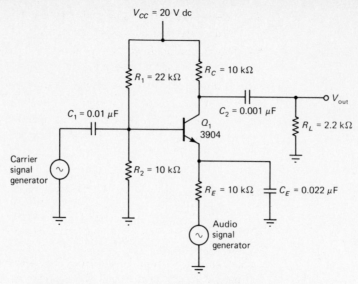

FIGURE 12-3   Class A, AM DSBFC modulator-emitter modulation

in the dc emitter current. Changes in the emitter current produce changes in the dynamic ac emitter resistance (r'_e), causing a corresponding change in the voltage gain of the amplifier. The resultant is an AM DSBFC envelope at the output.

Procedure

1. Construct the class A emitter modulator circuit shown in Figure 12-3.
2. Calculate the quiescent, maximum, and minimum voltage gains for the circuit for a modulation signal voltage $V_a = 8$ V p–p.
3. From the results of step 2, determine the quiescent, maximum, and minimum output voltages for a carrier input signal voltage $v_c = 0.1$ V p–p.
4. Measure the actual quiescent, maximum, and minimum voltage gains by setting the amplitude of the carrier signal generator output voltage to 0.1 V p–p and a frequency $F_c = 1$ MHz. Set the amplitude of the audio signal generator output voltage to 8 V p–p and a frequency $F_a = 1$ kHz.
5. Carefully adjust the output voltages from the two signal generators until a symmetrical AM envelope is observed at V_{out}.
6. Sketch the waveforms observed at the base, emitter, and collector of Q_1 and at V_{out}.
7. Describe the waveforms observed in step 6 in terms of frequency content, amplitude, and phase.
8. Determine the % modulation using the following formula.

$$\% \text{ mod} = \frac{V_{max} - V_{min}}{V_{max} + V_{min}} \times 100$$

where V_{max} = maximum peak-to-peak envelope voltage
V_{min} = minimum peak-to-peak envelope voltage

9. Increase the amplitude of the audio signal generator output voltage until 100% AM modulation is achieved.
10. Sketch the output waveform observed in step 9.
11. Decrease the amplitude of the audio signal generator output voltage until 50% AM modulation is achieved.
12. Sketch the output waveform observed in step 11.

13. Slowly vary first the amplitude and then the frequency of the audio signal generator while observing the output envelope.

14. Describe what effect varying the frequency and amplitude of the audio signal generator had on the output envelope.

15. Slowly change first the amplitude and then the frequency of the carrier signal generator while observing the output envelope.

16. Describe what effect varying the frequency and amplitude of the carrier signal generator has on the output envelope.

SECTION C   Trapezoidal Patterns

In this section a standard oscilloscope is used to display trapezoidal patterns. Trapezoidal patterns are often used for determining the percent modulation, linearity, and general shape of an AM envelope because they are easier to interpret than the envelope itself. The schematic diagram for the test circuit used to generate trapezoidal patterns is shown in Figure 12-4. On some oscilloscopes, the $X - Y$ mode is used for producing trapezoidal patterns. With the $X - Y$ mode, the audio signal is applied to the X input and the modulated envelope is applied to the Y input.

Procedure

1. Construct the test circuit shown in Figure 12-4.

2. Set the amplitude of the carrier signal generator output voltage to 0.1 V p–p and a frequency $F_c = 1$ MHz, and set the amplitude of the audio signal generator output voltage to 8 V p–p and a frequency $F_a = 1$ kHz.

3. Sketch the trapezoidal pattern observed on the oscilloscope.

4. Determine the percent modulation.

5. Adjust the amplitude of the audio signal generator output voltage until 100% AM modulation is achieved.

6. Sketch the waveform observed in step 5.

7. Reduce the amplitude of the audio signal generator output voltage until 50% AM modulation is achieved.

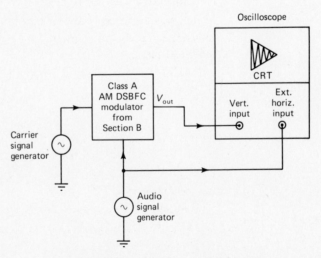

FIGURE 12-4 Trapezoidal pattern test set up

8. Sketch the waveform observed in step 7.

9. Increase the amplitude of the audio signal generator output voltage until a percent modulation greater than 100% is achieved.

10. Sketch the waveform observed in step 9.

11. Adjust the amplitude of the audio signal generator output voltage for 100% AM modulation.

12. Reverse the inputs to the horizontal (X) and vertical (Y) connectors of the oscilloscope.

13. Describe what effect reversing the horizontal and vertical inputs has on the trapezoidal pattern.

SECTION D   Summary

Write a brief summary of the concepts presented in this experiment on class A, AM DSBFC modulators. Include the following items:

1. The characteristics of class A common-emitter amplifiers.

2. The relationship between the amplitude and frequency of the modulating signal and the voltage gain of the amplifier.

3. The relationship between the amplitude and frequency of the modulating signal and the output AM envelope.

4. The relationship between a trapezoidal pattern and an AM envelope.

EXPERIMENT 12 Answer Sheet

NAME: _____ CLASS: _____ DATE: _____

SECTION A

2. A_v (calculated) = _____   3. A_v (measured) = _____

4. % error = _____

5.

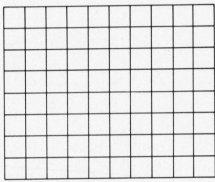

Vertical sensitivity _____ V/cm

Time base _____ sec/cm

Vertical sensitivity _____ V/cm

Time base _____ sec/cm

6. _____

_____

_____

SECTION B

2. A_q (calculated) = _____   3. V_q (calculated) = _____

A_{max} (calculated) = _____ V_{max} (calculated) = _____

A_{min} (calculated) = _____ V_{min} (calculated) = _____

4. A_q (measured) = _____

A_{max} (measured) = _____

A_{min} (measured) = _____

6.

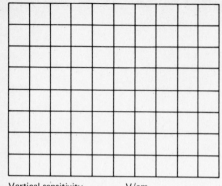

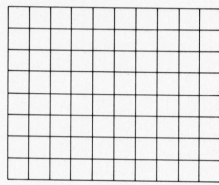

Vertical sensitivity _____ V/cm

Time base _____ sec/cm

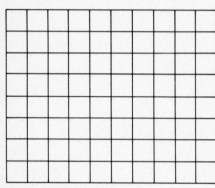

Vertical sensitivity _____ V/cm

Time base _____ sec/cm

Vertical sensitivity _____ V/cm

Time base _____ sec/cm

Vertical sensitivity _____ V/cm

Time base _____ sec/cm

7. _____

_____

_____

8. % modulation = _____

10.

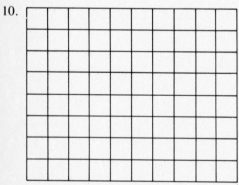

Vertical sensitivity _____ V/cm

Time base _____ sec/cm

12.

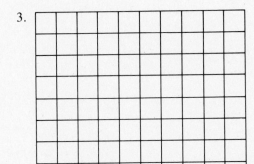

Vertical sensitivity _____ V/cm

Time base _____ sec/cm

14. _____

_____

_____

16. _____

_____

_____

SECTION C

3.

Vertical sensitivity _____ V/cm

Time base _____ sec/cm

4. % modulation = _____

6.

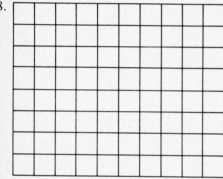

Vertical sensitivity _____ V/cm

Time base _____ sec/cm

8.

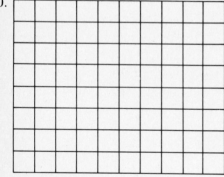

Vertical sensitivity _____ V/cm

Time base _____ sec/cm

10.

Vertical sensitivity _____ V/cm

Time base _____ sec/cm

13. _____

_____

_____

CLASS C, AM DSBFC MODULATOR

REFERENCE TEXT: *Fundamentals of Electronic Communications Systems.*

1. Chapter 3, Medium Power AM DSBFC Transistor Modulator.
2. Chapter 3, Trapezoidal Patterns.

OBJECTIVES

1. To observe the characteristics of a class C transistor amplifier.
2. To observe the operation of a simple AM DSBFC modulator using collector modulation.
3. To observe the operation of a class C transistor modulator.
4. To observe an AM DSBFC envelope and the relationship between the AM envelope, the modulating signal, and the carrier.

INTRODUCTION

Class A transistor modulators, such as the one used in Experiment 12, are inefficient and therefore impractical for medium- and high-power applications. However, class C transistor modulators have a maximum theoretical efficiency of 80% and are therefore used whenever power efficiency is a primary concern. In this exercise amplitude modulation is accomplished using a class C transistor modulator. A low-amplitude carrier signal is applied to the base of the transistor. With no modulating signal, the output is simply the input signal amplified and inverted. The modulating signal is applied to the collector, which adds to and subtracts from the dc supply voltage. This technique is called collector modulation. Unfortunately, a relatively high amplitude modulating signal voltage is required with collector modulation, creating an obvious disadvantage. The primary advantage of collector modulation is increased power efficiency. The block diagram for the test circuit used in this experiment is shown in Figure 13-1.

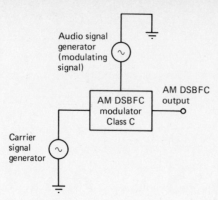

FIGURE 13-1 AM DSBFC modulator test circuit

MATERIALS REQUIRED

Equipment:

1 — protoboard

1 — dc power supply (15 V dc)

1 — audio signal generator

1 — medium-frequency signal generator

1 — standard oscilloscope (10 MHz)

1 — assortment of test leads and hookup wire

Parts List:

1 — small signal transistor (3904 or equivalent)

1 — 3-mH inductor

1 — 10 k-ohm resistor

1 — 100 k-ohm resistor

2 — 0.001-μF capacitors

1 — 1-μF capacitor

SECTION A   Class C Tuned Transistor Amplifier

In this section the characteristics of a class C tuned transistor amplifier are examined. Class C amplifiers conduct for less than 180° of the input signal. Therefore, collector current flows for only a small portion of the input signal and the output current waveform resembles a train of narrow pulses. The flywheel effect in the tuned collector circuit converts the narrow current pulses to a sinusoidal voltage waveform with an amplitude equal to approximately two times the collector supply voltage (i.e., 2 V_{CC}). Therefore, once the barrier potential of the base-emitter junction has been exceeded, the output voltage is independent of the ac base drive voltage. For a silicon transistor, a base drive of approximately 2 V p–p is sufficient to forward bias the transistor during the positive peaks of the input signal. The schematic diagram for the class C amplifier used in this section is shown in Figure 13-2.

Procedure

1. Construct the class C transistor amplifier circuit shown in Figure 13-2.
2. Calculate the resonant frequency of the tank circuit using the following formula.

$$F_r = \frac{1}{2\pi\sqrt{LC}}$$

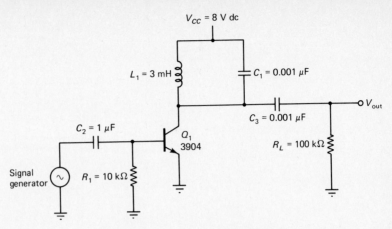

FIGURE 13-2   Class C tuned amplifier

where F_r = resonant frequency
$L = L_1$
$C = C_1$

3. Measure the actual resonant frequency by setting the amplitude of the signal genera-tor output voltage to 2 V p–p and varying its output frequency from 80 to 150 kHz. (It may be necessary to reduce the amplitude of the signal generator output voltage to produce an undistorted output wave with a peak-to-peak amplitude equal to ap-proximately 2 V_{CC}.)

4. Calculate the voltage gain for the amplifier using the following formula.

$$A_v = \frac{V_{out}}{V_{in}}$$

where A_v = voltage gain
V_{out} = ac output voltage ($\approx 2\ V_{CC}$)
V_{in} = ac base voltage (≈ 2 V p–p)

5. Determine the minimum input voltage by decreasing the amplitude of the signal generator output voltage until the amplifier output voltage (V_{out}) reduces signifi-cantly.

6. Determine the maximum peak-to-peak output voltage by varying the amplitude of the signal generator output voltage until V_{out} reaches its maximum undistorted value.

7. Increase the collector supply voltage (V_{CC}) to 12 V dc and repeat step 6.

8. Describe what effect increasing the dc supply voltage has on the amplifier output voltage.

SECTION B   Class C, AM DSBFC Modulator

In this section the characteristics of a class C, AM DSBFC transistor modulator are exam-ined. The schematic diagram for the modulator circuit used in this section is shown in Figure 13-3. Notice that the modulator circuit shown in Figure 13-3 is identical to the class C amplifier used in Section A except with the addition of the audio signal generator in the collector circuit. In this section a relatively high frequency carrier signal is applied

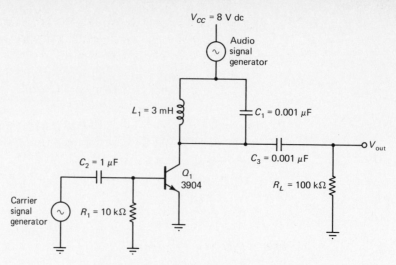

FIGURE 13-3 Class C, AM DSBFC modulator

to the base of Q_1, and a relatively low frequency modulating signal is applied to the collector. Thus, this circuit is a collector modulator. With collector modulation, the modulating signal adds to and subtracts from the dc supply voltage, producing corresponding changes in the amplitude of the output waveform. The output waveform is the sum of the unmodulated carrier and audio signal voltages (i.e., the peak-to-peak output voltage is approximately equal to 2 V_{CC} plus the peak audio voltage).

Procedure

1. Construct the class C collector modulator circuit shown in Figure 13-3.
2. Set the carrier signal generator output voltage to 2 V p–p and adjust its frequency to the value determined in step 3 of Section A.
3. With the output of the audio signal generator set to minimum, slowly vary the carrier signal generator output amplitude and frequency until V_{out} is at its maximum undistorted value.
4. Set the audio signal generator output frequency to 100 Hz and increase its amplitude until 100% AM modulation is observed at V_{out}. (The amplitude of the carrier signal generator output voltage may have to be varied slightly to produce an undistorted AM envelope.)
5. Sketch the waveforms for the audio and carrier input signals and the output AM envelope (V_{out}).
6. Measure the maximum and minimum peak-to-peak amplitudes of the output envelope.
7. Vary first the amplitude and then the frequency of the audio signal generator, and describe the relationship between the amplitude and frequency of the modulating signal and the AM envelope.
8. Increase the dc input voltage (V_{CC}) to 10 V dc.
9. Increase the amplitude of the audio signal generator output voltage until 100% AM modulation is observed at V_{out}.
10. Describe the relationship between the dc supply voltage, the modulating signal voltage, and the AM envelope.
11. Slowly vary the amplitude of the carrier signal generator output voltage. Describe what effect varying it has on the modulated envelope.
12. Set up the oscilloscope to display a trapezoidal pattern as follows: connect the exter-

nal horizontal input of the oscilloscope to the output of the audio signal generator, and the vertical input of the oscilloscope to the output of the modulator. (On some oscilloscopes, X and Y inputs are used rather than the vertical and external horizontal inputs.)

13. Adjust the amplitude of the audio signal generator output voltage for a trapezoidal pattern showing 100% AM modulation.

14. Sketch the waveform observed in step 13.

15. Vary the amplitude of the audio signal generator output voltage. Describe what effect varying it has on the trapezoidal p attern.

16. Vary the amplitude of the carrier signal generator. Describe what effect varying it has on the trapezoidal pattern.

SECTION C   Summary

Write a brief summary of the concepts presented in this experiment on class C, AM DSBFC modulators. Include the following items:

1. The characteristics of class C tuned amplifiers.

2. The relationship between the base input voltage and the collector output voltage in a class C tuned amplifier.

3. The relationship between the dc supply voltage and the ac collector voltage in a class C tuned amplifier.

4. The relationship between the modulating signal amplitude and frequency and the AM envelope in a class C, AM DSBFC modulator.

5. The relationship between the carrier amplitude and the AM envelope in a class C, AM DSBFC modulator.

EXPERIMENT 13   Answer Sheet

NAME: _____   CLASS: _____   DATE; _____

SECTION A

2. F_r (calculated) = _____   3. F_r (measured) = _____

4. A_v = _____   5. V_{in} (min) = _____

6. V_{out} (max) = _____   7. V_{out} (max) = _____

8. _____

   _____

   _____

SECTION B

5.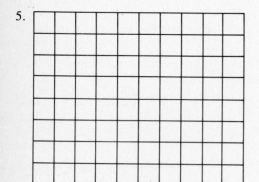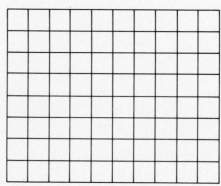

Vertical sensitivity _____ V/cm   Vertical sensitivity _____ V/cm

Time base _____ sec/cm   Time base _____ sec/cm

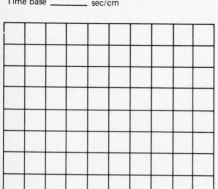

Vertical sensitivity _____ V/cm

Time base _____ sec/cm

6. V_{out} (max) = _____ V_{out} (min) = _____

7. _____

_____

_____

10. _____

_____

_____

11. _____

_____

_____

14.

Vertical sensitivity _____ V/cm

Time base _____ sec/cm

15. _____

_____

_____

16. _____

_____

_____

LINEAR INTEGRATED-CIRCUIT AM MODULATORS

REFERENCE TEXT: *Fundamentals of Electronic Communications Systems.*

1. Chapter 6, Linear Integrated-Circuit Balanced Modulators.
2. Chapter 2, Large-Scale-Integration Oscillators.
3. Chapter 5, Voltage-Controlled Oscillator.

OBJECTIVES

1. To observe the operation of a linear integrated-circuit function generator.
2. To observe the output amplitude-versus-input voltage characteristics of a linear integrated-circuit function generator.
3. To observe the operation of a linear integrated-circuit AM DSBFC modulator.
4. To observe the operation of a linear integrated-circuit AM DSBSC modulator.

INTRODUCTION

Linear integrated circuit (LIC) function generators are ideally suited for communications and instrumentation applications requiring a low output power, amplitude modulated signal. LIC AM modulators offer excellent frequency stability, linear amplitude modulation characteristics, circuit miniaturization, and simplicity of design. In this experiment the operation of the XR-2206 monolithic function generator as an AM modulator is examined. The block diagram for the XR-2206 function generator is shown in Figure 14-1. The XR-2206 function generator comprises four functional blocks: a voltage-controlled oscillator, an analog multiplier and sine shaper, a unity gain buffer amplifier, and a set of current switches.

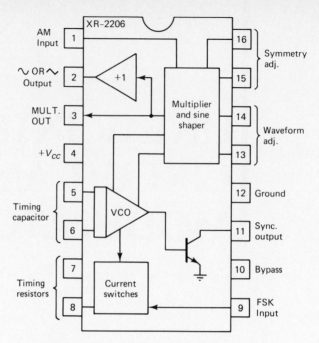

FIGURE 14-1 XR-2206 Functional block diagram

MATERIALS REQUIRED

Equipment:

1 — protoboard

1 — dual dc power supply (+ 12 V dc and 0 to + 5 V dc)

1 — audio signal generator

1 — standard oscilloscope (10 MHz)

1 — assortment of test leads and hookup wire

Note: If the audio signal generator is equipped with a variable dc offset, the 0 to + 5 V dc supply is not needed.

Parts List:

1— XR-2206 monolithic function generator

2 — 4.7 k-ohm resistors

1 — 10 k-ohm resistor

1 — 47 k-ohm resistor

1 — 1 k-ohm variable resistor

2 — 0.001-μF capacitors

2 — 1-μF capacitors

1 — 10-μF capacitor

SECTION A   Output Amplitude-versus-Input Bias Voltage

In this section the output amplitude-versus-input voltage characteristics of the XR-2206 linear integrated-circuit function generator are examined. The output amplitude of the XR-2206 function generator can be varied by applying a dc bias voltage to pin 1. The output amplitude varies linearly with the bias voltage for values within ±4 volts of $V_+/2$.

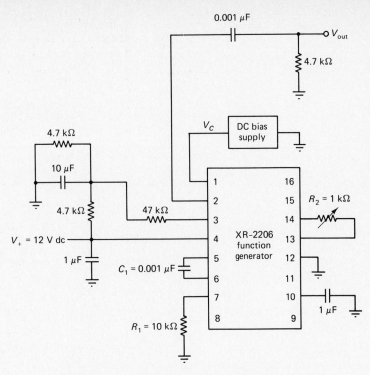

FIGURE 14-2   Output amplitude-versus-input bias voltage

The schematic diagram of the function generator circuit used in this section is shown in Figure 14-2.

Procedure

1. Construct the function generator circuit shown in Figure 14-2. Set the dc bias voltage (V_C) to 0 V dc.
2. Calculate the free-running oscillator frequency using the following formula.

$$F_o = \frac{1}{RC}$$

where F_o = free-running frequency
$R = R_1$
$C = C_1$

3. Vary R_2 until a sine wave with minimum distortion is observed at V_{out}.
4. Measure the frequency and amplitude of the sine wave observed in step 3.
5. Increase the dc bias voltage (V_C) in one-volt steps to +8 V dc. Measure the amplitude of the sine wave at V_{out} for each value of bias voltage.
6. Construct a graph of the output amplitude versus dc bias voltage for the bias voltages given and output amplitudes measured in step 5.
7. Describe the relationship between the dc bias voltage and the output frequency.

SECTION B   Linear Integrated-Circuit AM DSBFC Modulator

In this section the XR-2206 function generator is used to generate an AM DSBFC envelope. If the bias voltage applied to pin 1 of the XR-2206 contains both a dc bias and a sinusoidal component, the amplitude of the output signal on pin 2 will vary proportionately to the sinusoidal input signal. That is, the sinusoidal component of the bias signal

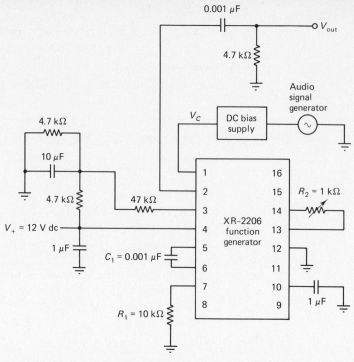

FIGURE 14-3 AM Modulator using the XR-2206 function generator

amplitude modulates the output signal to produce an AM DSBFC envelope. The schematic diagram for the linear integrated circuit AM DSBFC modulator used in this section is shown in Figure 14-3. The free-running frequency of the function generator is the carrier signal, and the audio signal generator output signal is the modulating (information) signal.

Procedure

1. Construct the linear integrated-circuit AM DSBFC modulator circuit shown in Figure 14-3.
2. Set the dc bias voltage (V_C) to 0 V dc.
3. Set the amplitude of the audio signal generator output voltage to 0 V and a frequency $F_a = 100$ Hz.
4. Calculate the free-running frequency of the function generator.
5. Vary R_2 until a sine wave with minimum distortion is observed at V_{out}.
6. Increase the dc bias voltage (V_C) to +5 V dc.
7. Increase the amplitude of the audio signal generator output voltage to 3 V p–p.
8. Adjust the amplitude of the audio signal generator output voltage until an AM envelope with 100% modulation is observed at V_{out}.
9. Sketch the waveform observed in step 8.
10. Describe the waveform sketched in step 9 in terms of frequency content, amplitude, shape, and repetition rate.
11. Vary the frequency and amplitude of the audio signal generator output voltage and describe what effect it has on the output waveform.
12. Set the amplitude of the audio signal generator output voltage to 1.5 V p–p and determine the percent modulation using the following formula.

$$\%M = \frac{V_{max} - V_{min}}{V_{max} + V_{min}} \times 100$$

where $\%M$ = percent modulation
V_{max} = maximum peak-to-peak voltage
V_{min} = minimum peak-to-peak voltage

13. Make the appropriate connections between the circuit and the oscilloscope and the necessary adjustments on the oscilloscope to display a trapezoidal pattern.

SECTION C   Linear Integrated-Circuit AM DSBSC Modulator

In this section the XR-2206 function generator is used to generate an AM DSBSC envelope. As shown in Section A, the functio n generator output amplitude varies linearly with the dc bias voltage applied to pin 1 for values of dc bias between ± 4 volts of V_+. As the dc bias approaches $V_+/2$, the phase of the output signal is reversed, and the amplitude goes through 0 (i.e., the free-running frequency of the function generator VCO is suppressed). This property of the function generator makes it suitable for suppressed carrier AM generation. The schematic diagram for the linear integrated-circuit AM DSBSC modulator circuit used in this section is identical to the AM DSBFC modulator circuit used in Section B and shown in Figure 14-3.

Procedure

1. Construct the linear integrated-circuit AM DSBSC modulator circuit shown in Figure 14-3.
2. Set the dc bias voltage (V_C) to 0 V dc.
3. Set the amplitude of the audio signal generator output voltage to 0 V and a frequency F_a = 100 Hz.
4. Vary R_2 until a sine wave with minimum distortion is observed at V_{out}.
5. Increase the dc bias voltage to 6 V dc (i.e., $V_+/2$).
6. Vary the dc bias voltage slightly above and below 6 V dc while observing V_{out}.
7. Describe the change in the output waveform observed in step 6.
8. Increase the amplitude of the audio signal generator output voltage until V_{out} is a symmetrical AM DSBSC envelope with minimum distortion. (It may be necessary to adjust the dc bias voltage slightly to achieve a symmetrical envelope.)
9. Sketch the waveform observed in step 8.
10. Describe the waveform sketched in step 9 in terms of frequency content, amplitude, shape, and repetition rate.
11. Vary the dc bias voltage slightly above and below 6 V dc while observing the phase of V_{out} (hint: use the external trigger).
12. Describe the change in the output waveform observed in step 11.
13. Adjust the dc bias until a symmetrical AM DSBSC envelope is observed at V_{out}.
14. Vary the frequency and amplitude of the audio signal generator output voltage and describe what effect it has on V_{out}.

SECTION D   Summary

Write a brief summary of the concepts presented in this experiment on linear integrated circuit AM modulators. Include the following items:

1. The relationship between the dc bias voltage and the amplitude of the function generator output voltage.

2. The relationship between the frequency and amplitude of the audio signal and the function generator output voltage for the AM DSBFC modulator.

3. The relationship between the frequency and amplitude of the audio signal and the function generator output voltage for the AM DSBSC modulator.

4. The relationship between the dc bias voltage and the phase of the function generator output signal.

EXPERIMENT 14   Answer Sheet

NAME: _____ CLASS: _____ DATE: _____

SECTION A

2. F_o (calculated) = _____   4. F_o (measured) = _____

                                                      V_o = _____

5.

V (V)	F_o	V_o	V (V)	F_o	V_o
1			5		
2			6		
3			7		
4			8		

6.

DC Bias Voltage

7. _____

_____

_____

SECTION B

4. F_o (calculated) = _____

9.

Vertical sensitivity _____ V/cm

Time base _____ sec/cm

10. _____

_____

_____

11. _____

_____

_____

12. % modulation = _____

SECTION C

7. _____

_____

_____

9.

Vertical sensitivity _____ V/cm

Time base _____ sec/cm

10. _____

_____

_____

12. _____

_____

_____

14. _____

_____

_____

TUNED AMPLIFIERS

REFERENCE TEXT: *Fundamentals of Electronic Communications Systems.*

1. Chapter 4, RF Amplifier Circuits.
2. Chapter 5, IF Amplifier Circuits.

OBJECTIVES

1. To observe the gain-versus-frequency response characteristics of a tuned class A transistor amplifier.
2. To observe the gain-versus-frequency response characteristics of a tuned class C transistor amplifier.
3. To determine the bandwidth and quality factor of an LC tank circuit.

INTRODUCTION

In electronic communications circuits it is often necessary to amplify a signal with a specific frequency or signals within a specific band of frequencies more than other signals. Tuned amplifiers have a frequency response curve similar to that of a bandpass filter except with gain. In radio communications receivers, tuned amplifiers are commonly used to separate a desired frequency or band of frequencies from a composite waveform. For example, broad-tuned radio frequency amplifiers are used to block undesired frequencies, such as the image frequency, from entering a radio receiver, and intermediate-frequency amplifiers separate and amplify a narrow band of signals for base-band detection. In this exercise the frequency response characteristics of both class A and class C tuned transistor amplifiers are examined. The block diagram of the test circuit used in this experiment is shown in Figure 15-1.

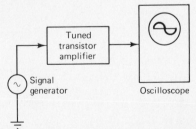

FIGURE 15-1 Tuned circuit block diagram

MATERIALS REQUIRED

Equipment:

1 — protoboard
1 — dc power supply (12 V dc)
1 — medium-frequency signal generator (1 MHz)
1 — standard oscilloscope (10 MHz)
1 — assortment of test leads and hookup wire

Parts List:

1 — small signal transistor (3904 or equivalent)
1 — 3-mH inductor
1 — 1 k-ohm resistor
1 — 10 k-ohm resistor
1 — 47 k-ohm resistor
2 —100 k-ohm resistors
2 — 1-μF capacitors
1 — 0.0001-μF capacitor

SECTION A   Class A Tuned Amplifier

In this section the gain-versus-frequency response characteristics of a class A tuned amplifier are examined. As previously discussed and demonstrated in Experiment 2, the impedance of a tank circuit is a function of frequency with a maximum value at resonance. A tank circuit, placed in the collector of a class A transistor amplifier, determines the frequency response characteristics of the amplifier. The ac collector resistance and consequently the voltage gain of the amplifier are maximum at the resonant frequency and decrease as frequency increases or decreases from resonance. In a class A amplifier, collector current flows for the entire 360° of the input signal. Therefore, due to the constant current drain, the maximum theoretical efficiency is only 25%. In a class A tuned amplifier, ac collector voltage is dependent on ac base drive voltage, although the voltage gain is relatively constant. However, the maximum collector output voltage is limited to twice the quiescent collector-to-emitter voltage (i.e., $v_c = 2\,V_{CE_q}$). The schematic diagram for the class A tuned amplifier circuit used in this section is shown in Figure 15-2.

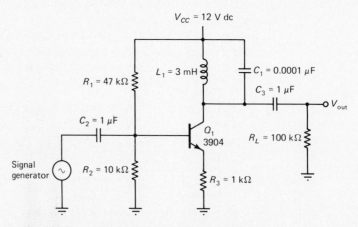

FIGURE 15-2   Class A tuned amplifier

Procedure

1. Construct the class A tuned amplifier circuit shown in Figure 15-2.

2. Calculate the resonant frequency of the tank circuit using the following formula.

$$F_r = \frac{1}{2\pi\sqrt{LC}}$$

where F_r = resonant frequency
$C = C_1$
$L = L_1$

3. Measure the actual resonant frequency by setting the signal generator output voltage to 100 mV p–p and varying its output frequency from 100 kHz to 1 MHz.

4. Plot the gain-versus-frequency response curve for the tuned amplifier on semilog graph paper.

5. Calculate the percentage error between the calculated and actual resonant frequencies using the following formula.

$$\% \text{ error} = \frac{\text{measured value} - \text{calculated value}}{\text{calculated value}} \times 100$$

6. Measure the voltage gain of the amplifier at the resonant frequency using the following formula.

$$A_v = \frac{V_{\text{out}}}{V_{\text{in}}} = \frac{v_c}{v_b}$$

where A_v = voltage gain
V_{out} = ac output voltage
V_{in} = ac base voltage

7. From the frequency response curve, determine the upper and lower break frequencies.

8. What is the -3-dB bandwidth of the amplifier?

9. Determine the quality (Q) factor for the tank circuit using the following formula.

$$Q = \frac{F_r}{B}$$

where Q = Q factor
$B = -3$-dB bandwidth
F_r = resonant frequency

10. Measure the quiescent dc voltage on the emitter, base, and collector of Q_1, and calculate the quiescent collector-to-emitter voltage.

11. From the measurements made in step 10, determine the maximum ac output voltage at the collector of Q_1.

12. Determine the maximum peak-to-peak output voltage by increasing the amplitude of the signal generator output until V_{out} reaches its maximum undistorted value.

13. Measure the voltage gain with the output voltage at maximum and compare the value to the gain measured in step 6.

14. Design a class A tuned amplifier with a resonant frequency $F_r = 134$ kHz. Use an inductance $L = 3$ mH.

15. Construct the tuned amplifier circuit designed in step 14, and plot its gain-versus-frequency response curve on semilog graph paper.

SECTION B   Class C Tuned Amplifier

In this section the frequency response characteristics of a tuned class C transistor amplifier
are examined. Class C amplifiers conduct for less than 180° of the input signal, and the
collector current waveform resembles a train of narrow pulses. Therefore, collector current flows for a small portion of the input signal, and class C amplifiers have a maximum
theoretical efficiency of nearly 100%. Also, because of the flywheel effect, the tank circuit converts the narrow current pulses to a sinusoidal voltage wave with an amplitude
equal to two times the collector supply voltage (i.e., 2 V_{CC}). Therefore, the collector output voltage from a tuned class C amplifier is independent of the ac base drive voltage once
the barrier potential of the base-emitter junction is exceeded. For a silicon transistor, a
base drive voltage of approximately 2 V p–p is sufficient to forward bias the transistor
during the positive peak of the input signal. The schematic diagram for the tuned class C
amplifier circuit used in this experiment is shown in Figure 15-3.

Procedure

1. Construct the class C tuned amplifier circuit shown in Figure 15-3.
2. Calculate the resonant frequency of the tank circuit.
3. Measure the actual resonant frequency by setting the signal generator output voltage
   to 2 V p–p and varying its output frequency from 80 to 150 kHz.
4. Calculate the percentage error between the calculated and actual resonant frequencies.
5. Plot the gain-versus-frequency response curve for the tuned amplifier on semilog
   graph paper.
6. Measure the quiescent dc voltage on the emitter, base, and collector of Q_1, and
   calculate the quiescent collector-to-emitter voltage.
7. Calculate the maximum ac output voltage at the collector of Q_1.
8. Measure the minimum ac input voltage by decreasing the amplitude of the signal
   output voltage until $V_{out} = 0$ V.
9. Measure the maximum ac collector output voltage by increasing the amplitude of
   the signal generator output voltage until V_{out} reaches its maximum undistorted amplitude.
10. Vary the collector supply voltage (V_{CC}) while observing V_{out}.
11. What effect does varying V_{CC} have on the ac output amplitude?

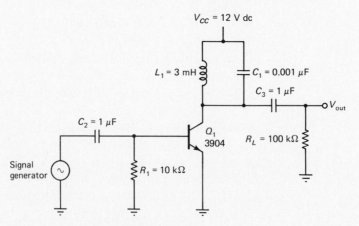

FIGURE 15-3   Class C tuned amplifier

12. Measure the maximum voltage gain of the amplifier by setting the amplitude of the signal generator output voltage to the value determined in step 8 and measuring the output voltage. Use the following formula.

$$A_v = \frac{V_{out}}{V_{in}}$$

where A_v = voltage gain
V_{out} = ac output voltage
V_{in} = minimum ac input voltage

13. Design a class C tuned amplifier with a resonant frequency F_r = 65 kHz (use an inductance L = 3 mH and a dc supply voltage V_{CC} = 10 V dc).

14. Construct the tuned amplifier circuit designed in step 13, and plot its gain-versus-frequency response curve on semilog graph paper.

15. From the frequency response curve, determine the upper and lower break frequencies.

16. What is the −3-dB bandwidth?

SECTION C   Summary

Write a brief summary of the concepts presented in this experiment on tuned amplifiers. Include the following items:

1. The gain-versus-frequency response characteristics of tuned class A transistor amplifiers.
2. The gain-versus-frequency response characteristics of tuned class C transistor amplifiers.
3. The relationship between resonant frequency, Q factor, and bandwidth for LC tuned circuits.
4. The relationship between voltage gain and input signal voltage for class A and class C tuned amplifiers.
5. The relationship between dc supply voltage and output amplitude for tuned class C amplifiers.

EXPERIMENT 15   Answer Sheet

NAME: _____ CLASS: _____ DATE: _____

SECTION A

2. F_r (calculated) = _____   3. F_r (measured) = _____

5. % error = _____   6. A_v = _____

7. F_U = _____ F_L = _____   8. B = _____

9. Q = _____   10. V_C = _____ V_B = _____

　　　　　　　　　　　　　　　　　　　　　　　V_E = _____ V_{CE} = _____

11. v_c (max) = _____   12. V_{p-p} (max) = _____

13. A_V = _____   14. C = _____

SECTION B

2. F_r (calculated) = _____   3. F_r (measured) = _____

4. % error = _____   6. V_C = _____ V_B = _____

　　　　　　　　　　　　　　　　　　　　　　　V_E = _____ V_{CE} = _____

7. v_c (max) = _____   8. v_b (min) = _____

9. v_c (max) = _____

11. _____

_____

_____

12. A_v (max) = _____   13. C = _____

15. F_U = _____ F_L = _____   16. B = _____

Experiment 16

AM PEAK DETECTOR

REFERENCE TEXT: *Fundamentals of Electronic Communications Systems.*

1. Chapter 4, AM Detector Circuits, Peak Detectors.
2. Chapter 4, AM Detector Circuits, Detector Distortion.

OBJECTIVES

1. To observe the operation of an AM peak detector.
2. To observe the relationship between the frequency and the amplitude of the carrier signal and the detected waveform.
3. To observe the relationship between the frequency and the amplitude of the audio signal and the detected waveform.
4. To observe the relationship between the AM envelope and the detected waveform.
5. To observe diagonal clipping and rectifier distortion.

INTRODUCTION

The simplest circuit for demodulating an AM DSBFC envelope is a peak detector. A peak detector is a noncoherent AM demodulator. It detects the peaks of the AM waveform and thus produces an output voltage with the shape of either the positive or negative half of the envelope. Therefore, the output voltage from a peak detector is proportional to both the amplitude and the frequency of the modulating signal. Essentially, the carrier component of the AM envelope "captures" the diode and forces it to turn on and off (rectify) synchronously. A lowpass filter, placed after the diode, separates the demodulated audio signal from the rectified AM envelope. The block diagram for the test circuit used in this experiment is shown in Figure 16-1.

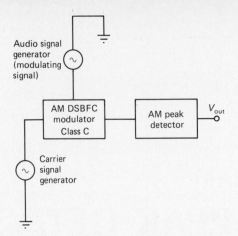

FIGURE 16-1 AM Peak detector test circuit

MATERIALS REQUIRED

Equipment:

1 — protoboard

1 — dc power supply (15 V dc)

1 — audio signal generator

1 — medium-frequency signal generator (1 MHz)

1 — standard oscilloscope (10 MHz)

1 — assortment of test leads and hookup wire

Parts List:

1 — small-signal transistor (3904 or equivalent)

1 — small-signal diode (1N914 or equivalent)

1 — 3-mH inductor

1 — 1 k-ohm resistor

2 — 10 k-ohm resistors

1 — 100 k-ohm resistor

1 — 1 M-ohm resistor

2 — 0.001-μF capacitors

1 — 0.01-μF capacitor

1 — 0.1-μF capacitor

1 — 1-μF capacitor

SECTION A   Peak Detector

In this section the rectifying action of a peak detector is examined. The schematic diagram for the peak detector used in this section is shown in Figure 16-2. During the positive half-cycle of the signal generator output waveform, D_1 conducts, charging C_1 with the polarity shown. During the negative half-cycle, D_1 is off, and C_1 discharges slightly through resistor R_1. Therefore, the output waveform is simply a half-wave rectified and filtered signal (i.e., a positive dc voltage).

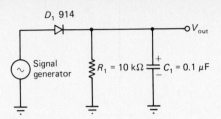

FIGURE 16-2 Peak detector schematic diagram

Procedure

1. Construct the peak detector circuit shown in Figure 16-2.
2. Set the signal generator output voltage to 4 V p–p and a frequency of 100 kHz.
3. Describe the output waveform in terms of its ac and dc components.
4. Vary the amplitude of the signal generator output voltage while observing the waveform at the output of the peak detector.
5. Describe what effect varying the amplitude of the signal generator output voltage had on the output waveform.
6. Slowly decrease the signal generator output frequency to 0 Hz while observing the waveform at the output of the peak detector.
7. Describe what effect varying the signal generator output frequency has on the output waveform.
8. Reverse the polarity of D_1 and repeat steps 2 through 5.
9. Describe any differences in the output waveform with the diode reversed.

SECTION B   AM Peak Detector

In this section the operation of an AM peak detector is examined. The input waveform to an AM peak detector comprises a carrier frequency and its upper and lower side frequencies (i.e., an AM envelope). A diode is a nonlinear device. Therefore, nonlinear mixing (heterodyning) occurs between the carrier and its associated side frequencies. The difference between the upper side frequency and the carrier frequency or the carrier frequency and the lower side frequency is the modulating signal (audio). The audio is separated from the composite waveform with a simple lowpass filter. The schematic diagram for the peak detector used in this section is shown in Figure 16-3. The circuit comprises a class C, AM modulator and a peak detector. During the positive half-cycles of the AM envelope, D_1 is forward biased (on), and thus, C_1 charges to a value equal to the peak envelope voltage minus the diode barrier potential. During the negative half-cycles, D_1 is reverse biased (off), and C_1 discharges slightly through R_1. Each successive peak of an AM envelope has a different peak amplitude. Therefore, C_1 charges and discharges in proportion to the shape of the envelope (i.e., proportional to the modulating signal). Thus, the waveform on the input side of the diode is a clipped envelope, and the waveform on the output side of the diode is the detected audio signal.

Procedure

1. Construct the AM modulator and peak detector circuits shown in Figure 16-3.
2. Set the amplitude of the carrier signal generator output voltage to 2 V p–p and adjust its frequency to the resonant frequency of the tank circuit (L_1 and C_1).
3. Set the amplitude of the audio signal generator output voltage to 6 V p–p and adjust its frequency to 100 Hz.

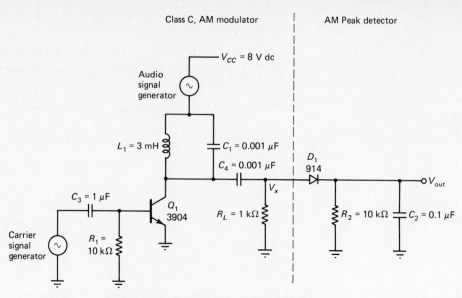

FIGURE 16-3 AM DSBFC Demodulator-peak detector

4. Slowly vary the amplitude and frequency of the two signal generators until a 50% modulated AM envelope is observed at V_x.

5. Sketch the waveforms observed at V_x and V_{out}.

6. Describe the waveforms observed in step 4 in terms of peak amplitude, frequency content, and ac and dc voltages.

7. Vary the frequency and amplitude of the audio signal generator while observing V_{out}.

8. Describe what effect varying the frequency and amplitude of the audio signal generator has on the output waveform.

9. Slowly vary the voltage of the collector supply (V_{CC}) between 6 and 12 V dc while observing V_x and the output waveform.

10. Describe what effect varying the dc supply voltage has on the waveforms observed in step 9. (Include the effect on both the ac and dc components.)

11. Replace R_2 with a 1 M-ohm resistor.

12. Sketch the waveform observed at V_{out}.

13. Name the distortion present in the output waveform.

14. Describe the waveform sketched in step 12.

15. Replace R_2 with the original 10 k-ohm resistor, and replace C_2 with a 0.001-μF capacitor.

16. Sketch the waveform observed at V_{out}.

17. Name the distortion present in the output waveform.

18. Describe the waveform sketched in step 16.

SECTION C   Summary

Write a brief summary of the concepts presented in this experiment on AM peak detectors. Include the following items:

1. The relationship between the frequency and amplitude of the carrier and the detected waveform.

2. The relationship between the frequency and amplitude of the audio and the detected waveform.

3. The relationship between the AM envelope and the detected waveform.

4. The relationship between the dc supply voltage and the ac and dc components of the detected waveform.

5. The causes and effects of diagonal clipping and rectifier distortion.

EXPERIMENT 16   Answer Sheet

NAME: _____ CLASS: _____ DATE: _____

SECTION A

3. _____
_____
_____

5. _____
_____
_____

7. _____
_____
_____

8. _____
_____
_____
_____
_____

9. _____
_____
_____

SECTION B

5.

Vertical sensitivity _____ V/cm

Time base _____ sec/cm

Vertical sensitivity _____ V/cm

Time base _____ sec/cm

6. _____

_____

_____

8. _____

_____

_____

10. _____

_____

_____

12.

Vertical sensitivity _____ V/cm

Time base _____ sec/cm

13. _____

14. _____

_____

_____

16.

Vertical sensitivity _____ V/cm

Time base _____ sec/cm

17. _____

18. _____

_____

_____

AUTOMATIC GAIN CONTROL (AGC)

REFERENCE TEXT: *Fundamentals of Electronic Communications Systems.*
Chapter 4, Automatic Gain Control.

OBJECTIVES

1. To observe the operation of an automatic gain control circuit.
2. To observe simple AGC.
3. To observe delayed AGC.

INTRODUCTION

In electronic communications receivers, it is often necessary to automatically adjust the gain of the receiver to compensate for changes in the received signal level. High-amplitude received signals can saturate the RF or IF amplifiers, producing harmonic and intermodulation distortion. Extremely weak received signals cannot provide the drive necessary to produce a usable signal at the output of an audio detector. An automatic gain-control circuit compensates for minor variations in the received signal level by supplying a feedback signal that automatically changes the IF and RF gains of the receiver. AGC is a form of degenerative or negative feedback (i.e., the AGC feedback signal reduces the receiver gain for strong received signals and increases the receiver gain for weak received signals). The two most common types of AGC are direct or simple AGC and delayed AGC. The block diagram for the test circuit used in this experiment is shown in Figure 17-1.

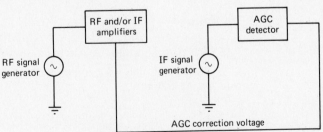

FIGURE 17-1 AGC test circuit

MATERIALS REQUIRED

Equipment:

1 — protoboard
1 — dc power supply (25 V dc)
2 — medium-frequency signal generators (1 MHz)
1 — standard oscilloscope (10 MHz)
1 — assortment of test leads and hookup wire

Parts List:

1 — small-signal transistor (3904 or equivalent)
1 — small-signal diode (914 or equivalent)
1 — 1 k-ohm resistor
1 — 4.7 k-ohm resistor
1 — 6.8 k-ohm resistor
3 — 10 k-ohm resistors
1 — 33 k-ohm resistor
1 — 100 k-ohm resistor
1 — 10 k-ohm variable resistor
2 — 0.001-μF capacitors
1 — 0.1-μF capacitor
2 — 0.5-μF capacitors

SECTION A   Direct or Simple AGC

In this section direct or simple AGC is examined. With simple AGC, the receiver gain is reduced whenever a detectable carrier signal is received. Therefore, degenerative feedback is immediate and reduces the receiver gain even for small received signals (i.e., AGC action occurs even for very weak received signals). This is the predominant disadvantage of simple AGC; it desensitizes the receiver. The primary advantage of simple AGC is its simplicity. Figure 17-2 shows the schematic diagram for the simple AGC circuit and common-emitter transistor amplifier used in this section. The AGC circuit is simply a peak detector. The IF signal generator simulates the IF carrier input to the AGC detector. The common-emitter amplifier simulates an RF or IF amplifier. The output signal from the peak detector is a negative dc voltage, which is fed back to the amplifier circuit where it combines with the quiescent dc bias and automatically adjusts the amplifier's voltage gain. The voltage gain of the amplifier is inversely proportional to the magnitude of the negative dc voltage. Therefore, increases in the signal amplitude at the input to the peak detector reduce the gain of the amplifier, and vice versa.

Procedure

1. Construct the simple AGC circuit shown in Figure 17-2.
2. Calculate the quiescent voltage gain for the common-emitter amplifier using the following formulas.

$$A_v = \frac{r_c}{r'_e}$$

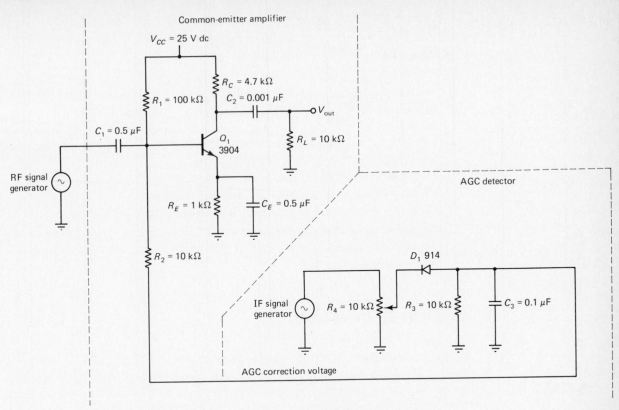

FIGURE 17-2 Simple AGC

where A_v = quiescent voltage gain
r_c = collector ac resistance
r'_e = dynamic ac emitter-base resistance

and
$$r'_e = \frac{25\ mV}{I_E} \qquad I_E = \frac{V_E}{R_E} \qquad r_c = \frac{R_C \times R_L}{R_C + R_L}$$

where V_E is a measured value and R_E = 1 k ohm

3. Set the amplitude of the IF signal-generator output voltage to 10 V p–p and a frequency F_{if} = 100 kHz.

4. Set R_4 to its minimum resistance value (i.e., 0 V ac at the cathode of diode D_1).

5. Measure the actual voltage gain of the amplifier as follows. Set the amplitude of the RF signal generator output voltage to 50 mV p–p and a frequency F_{rf} = 1 MHz. Measure the ac output voltage and determine the voltage gain using the following formula

$$A_v = \frac{V_{out}}{V_{in}}$$

where A_v = voltage gain
V_{out} = peak-to-peak ac output voltage
V_{in} = ac base voltage (50 mV p–p)

6. Calculate the percentage error between the calculated and actual voltage gains using the following formula.

$$\% \ error = \frac{measured\ value\ -\ calculated\ value}{calculated\ value} \times 100$$

7. Explain any difference between the calculated and actual voltage gains.
8. Record the dc voltage measured on the anode of diode D_1.
9. Adjust R_4 for a 3 Vp–p signal on the cathode of D_1.
10. Record the dc voltage on the anode of diode D_1.
11. Measure the voltage gain of the amplifier using the procedure outlined in step 5.
12. Increase the resistance of R_4 until the amplitude of V_{out} drops to an insignificant value (i.e., approximately 0 V). Record the ac voltage on the cathode of D_1, the voltage gain of the amplifier, and the dc voltage on the anode of D_1 after each adjustment of R_4.
13. Describe the relationship between the amplitude of the signal at the cathode of D_1, the dc voltage on the anode of D_1, and the voltage gain of the amplifier.
14. Set R_4 for minimum resistance (i.e., 0 V ac on the cathode of D_1).
15. Determine the IF input voltage where AGC action begins ($V_{threshold}$) by slowly increasing the resistance of R_4 until V_{out} just begins to decrease. Record the ac voltage on the cathode of D_1.

SECTION B   Delayed AGC

In this section delayed AGC is examined. With delayed AGC, the gain of a receiver is unaffected until the received signal level exceeds a predetermined amplitude. Therefore, the receiver gain is unaffected by weak received signals and reduced for strong received signals (i.e., AGC action is delayed until a significantly large signal is received). The schematic diagram for the delayed AGC circuit used in this section is shown in Figure 17-3. As in Section A, Q_1 is a common-emitter amplifier that simulates an RF or an IF

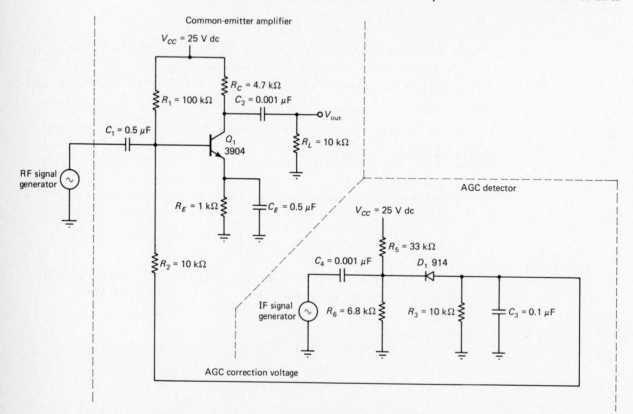

FIGURE 17-3   Delayed AGC

amplifier. The AGC circuit is again a peak detector. However, the cathode of D_1 is reverse biased by voltage divider R_5 and R_6 and by V_{CC}. Therefore, AGC action does not begin until the ac input voltage from the IF signal generator is sufficiently large to forward bias D_1.

Procedure

1. Construct the delayed AGC circuit shown in Figure 17-3.
2. Set the amplitude of the IF signal generator output voltage to 0 V and a frequency $F_{if} = 100$ kHz.
3. Set the amplitude of the RF signal generator output voltage to 50 mV p–p and a frequency $F_{rf} = 1$ MHz.
4. Increase the amplitude of the IF signal-generator output voltage in one-volt steps while observing both V_{out} and the dc voltage on the anode of D_1. Record the IF signal generator output voltage, the dc voltage on the anode of D_1, and the voltage gain of the amplifier after each one-volt step.
5. Describe the relationship between the ac voltage on the cathode of D_1, the dc voltage on the anode of D_1, and the voltage gain of the amplifier.
6. Calculate the IF signal voltage where AGC action begins (i.e., the delayed AGC threshold voltage) using the following procedure. Set the amplitude of the IF signal generator output voltage to 0 V and measure the dc voltage on the cathode and anode of D_1. Determine the peak amplitude of the IF threshold voltage using the following formula.

$$V_{th} = V_C - V_A + 0.7$$

where
$\quad V_{th} = $ peak IF threshold voltage
$\quad V_C = $ dc cathode voltage
$\quad V_A = $ dc anode voltage
$\quad 0.7\ V = $ diode barrier potential

7. Measure the threshold voltage by slowly increasing the amplitude of the IF signal generator output voltage until V_{out} just begins to decrease.
8. Compare the threshold voltage measured in step 7 to the voltage calculated in step 6.

SECTION C   Summary

Write a brief summary of the concepts presented in this experiment on automatic gain control. Include the following items:

1. The concept of simple AGC.
2. The concept of delayed AGC.
3. The relationship between detector input voltage, the AGC voltage, and the voltage gain of the IF and RF amplifiers for both simple and delayed AGC.
4. The concept of desensitizing a receiver.

EXPERIMENT 17 Answer Sheet

NAME: _____ CLASS: _____ DATE: _____

SECTION A

2. A_v (calculated) = _____

5. A_v (measured) = _____

6. % error = _____

7. _____

_____

_____

8. anode ac = _____

10. anode dc = _____

11. A_v = _____

12.

Cathode ac	Anode ac	Anode dc	A_v

13. _____

_____

_____

15. V_{th} = _____

SECTION B

4.

Cathode ac	Anode ac	Anode dc	A_v

5. _____

_____

_____

6. V_{th} (calculated) = _____

7. V_{th} (measured) = _____

8. _____

_____

_____

LINEAR INTEGRATED-CIRCUIT PHASE-LOCKED LOOP

REFERENCE TEXT: *Fundamentals of Electronic Communications Systems.*

1. Chapter 2, Large-Scale-Integration Oscillators.
2. Chapter 5, Phase-Locked Loop.

OBJECTIVES

1. To observe the operation of a voltage-controlled oscillator.
2. To observe the operation of a phase-locked loop.
3. To observe the operation of a linear integrated-circuit phase-locked loop.

INTRODUCTION

A phase-locked loop (PLL) is a closed-loop feedback control system in which the feedback signal is a frequency rather than simply a dc voltage. In this experiment the operation of the XR-2212 phase-locked loop is examined. The XR-2212 is an ultrastable monolithic phase-locked loop system especially designed for data communications and control-system applications. The XR-2212 PLL offers a 20 ppm temperature stability and is ideally suited for frequency synthesis, FM detection, FSK demodulation, and other tracking filter applications. The block diagram for the XR-2212 PLL is shown in Figure 18-1. The XR-2212 PLL comprises four primary sections: an input preamplifier, a phase detector, a stable voltage-controlled oscillator (VCO), and a high-gain differential amplifier. The VCO is brought out externally so that the circuit can operate as a frequency synthesizer using an external divider. The differential amplifier section can be used as an audio preamplifier for FM detection, or as a high-speed sense amplifier (comparator) for FSK demodulation. The center frequency, bandwidth, and tracking range of the PLL are controlled independently by the choice of external components. In essence, The VCO performs voltage-to-frequency conversion. The phase detector (or phase comparator as it is sometimes called) is a balanced mixer with two input frequencies: an external input frequency (F_i) and the VCO output frequency (F_o). Once lock has occurred, the phase comparator output signal is a dc voltage that is proportional to the difference in phase between F_o and F_i. The basic operation of a PLL is quite simple. The external input signal (F_i) is

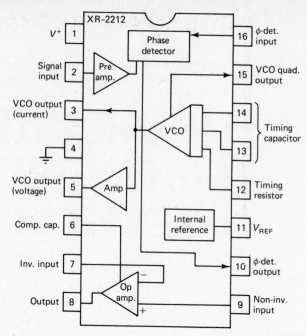

FIGURE 18-1 XR-2212 Phase-locked loop functional block diagram

compared to the VCO output signal (F_o) in the phase comparator. The dc output voltage from the phase comparator is fed back to the input of the VCO, where it deviates (shifts) the frequency of the VCO until its output frequency equals F_i. This process is called loop acquisition. Once the VCO output frequency equals F_i, the loop is said to be locked and the VCO output frequency will follow changes in the external input frequency.

MATERIALS REQUIRED

Equipment:

1 — protoboard
1 — dc power supply (+12 V dc)
1 — medium-frequency signal generator (1 MHz)
1 — standard oscilloscope (10 MHz)
1 — assortment of test leads and hookup wire

Parts List:

1 — XR-2212 monolithic phase-locked loop
1 — 1 k-ohm resistor
2 — 10 k-ohm resistors
1 — 22 k-ohm resistor
1 — 100 k-ohm resistor
1 — 0.001-μF capacitor
1 — 0.05-μF capacitor
2 — 0.1-μF capacitors

SECTION A Voltage-Controlled Oscillator

In this section the operation of the voltage-controlled oscillator in the XR-2212 is examined. A VCO is a free-running oscillator with a stable frequency of oscillation that is dependent on a timing capacitance, a timing resistance, and an external control voltage. The output from a VCO is a frequency, and its input is a bias or control voltage that can be either a dc or an ac voltage. In essence, a VCO performs voltage-to-frequency conversion. The schematic diagram for the VCO circuit used in this section is shown in Figure 18-2. The VCO free-running frequency (F_o) is determined by an external timing resistor (R_0) connected from pin 12 to ground and an external timing capacitance (C_0) connected between pins 13 and 14. The VCO free-running frequency can be measured on pin 5 with feedback resistor R_1 disconnected. The VCO output frequency can be varied by applying an external control voltage to the VCO input. The VCO will operate over a frequency range from 0.01 Hz to 300 kHz.

Procedure

1. Construct the PLL circuit shown in Figure 18-2.
2. Calculate the VCO free-running frequency using the following formula.

$$F_o = \frac{1}{RC}$$

where F_o = free-running frequency
$R = R_0$
$C = C_0$

3. Disconnect one end of resistor R_1 and measure the VCO free-running frequency on pin 5.
4. Calculate the percentage error between the calculated and actual free-running frequencies using the following formula.

$$\% \text{ error} = \frac{\text{measured value} - \text{calculated value}}{\text{calculated value}} \times 100$$

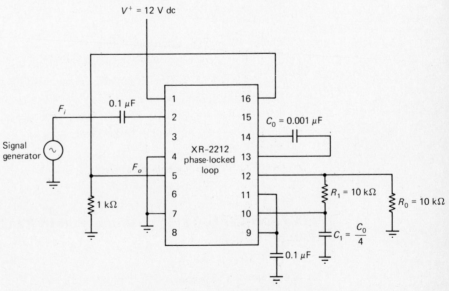

FIGURE 18-2 SR-2212 Phase-locked loop

5. Replace R_0 with a 22 k-ohm resistor and repeat steps 2, 3, and 4.
6. Replace C_0 with a 0.01-μF capacitor and repeat steps 2, 3, and 4.
7. Describe the relationship between the timing resistance (R_0), the timing capacitance (C_0), and the VCO free-running frequency.

SECTION B   Phase Comparator

In this section the operation of the phase comparator in the XR-2212 PLL is examined. The phase comparator is a balanced mixer with two inputs: an external input frequency (F_i) and the VCO output frequency (F_o). The phase comparator output is the product of F_o and F_i and, therefore, contains the sum and difference frequencies ($F_o + F_i$ and $F_o - F_i$). When $F_o = F_i$, the difference frequency is equal to 0 Hz (i.e., a dc voltage). When the external input frequency (F_i) is out of lock range or when there is no phase error within the PLL (i.e., $F_i = F_o$), the dc voltage at the phase comparator output (pin 10) is very nearly equal to the PLL reference voltage (V_r). V_r is the PLL's internal reference voltage for pins 10, 12, and 16. V_r appears on pin 11. The phase-comparator output voltage (V_d) is proportional to the difference in frequency between F_o and F_i. Therefore, as the external input frequency (F_i) deviates above and below F_o, V_d varies above and below V_r. The schematic diagram of the PLL circuit used in Section A and shown in Figure 18-2 is also used in this section.

Procedure

1. Construct the PLL circuit shown in Figure 18-2 with $R_0 = 10$ k ohms and $C_0 = 0.001$ μF.
2. Disconnect one end of R_1 and measure the VCO free-running frequency (F_o).
3. Reconnect resistor R_1.
4. Set the amplitude of the signal-generator output voltage to 4 V p–p and adjust its frequency to the VCO free-running frequency (i.e., $F_i = F_o$).
5. Calculate the PLL reference voltage using the following formula.

$$V_r = \frac{V_+}{2} - 0.65 \text{ V}$$

where V_r = PLL reference voltage
       V_+ = PLL dc supply voltage

6. Measure the PLL internal reference voltage (V_r) on pin 11.
7. Measure the phase comparator dc output voltage (V_d) on pin 10 and compare it to the reference voltage calculated in step 5 and measured in step 6.
8. Slowly vary the frequency of the external signal generator above and below F_o while observing V_d.
9. Describe the relationship between F_i, F_o, and V_d.

SECTION C   Loop Acquisition, Pull-In Range, and Hold-In Range

In this section the loop acquisition, pull-in range, and hold-in range for the XR-2212 PLL are examined. Essentially, loop acquisition is the process the PLL undergoes when it detects and locks onto an external input frequency. The PLL is said to be locked when the

VCO output frequency equals the external input frequency. The pull-in range is the peak range of input frequencies which the PLL will lock onto. The peak-to-peak pull-in range is often called the capture range. The lowest frequency at which lock occurs is called the lower capture frequency, and the highest frequency at which lock occurs is called the upper capture frequency. Hold-in range is the peak range of input frequencies over which the PLL will stay locked onto the input frequency once lock has occurred. The peak-to-peak hold-in range is often called the lock range. The lowest frequency that the PLL will track once lock has occurred is called the lower lock limit, and the highest frequency that the PLL will track once lock has occurred is called the upper lock limit. The difference between the lower lock limit and the upper lock limit is called the loop tracking bandwidth. The schematic diagram of the PLL circuit used in this section is the same as the circuit used in Sections A and B and shown in Figure 18-2.

Procedure

1. Construct the PLL circuit shown in Figure 18-2 with $R_0 = 10$ k ohms and $C_0 = 0.001$ μF.
2. Disconnect one end of R_1 and measure the VCO free-running frequency.
3. Reconnect R_1 and set the amplitude of the signal generator output voltage to 4 V p–p and adjust its frequency to 80 kHz.
4. Slowly increase the signal generator output frequency while observing both the VCO output signal and the signal generator output signal. Note that as F_i approaches F_o, the PLL locks onto the signal generator frequency and follows (tracks) it. However, if F_i is increased significantly beyond F_o, the PLL loses lock. After the PLL loses lock, slowly decrease the signal generator output frequency. Note that as F_i approaches F_o, the PLL locks onto the signal generator frequency and follows it. Again, if F_i is decreased significantly below F_o, the PLL loses lock.
5. Measure the lower capture frequency (F_{CL}) as follows: Set the signal generator output frequency to 50 kHz. Slowly increase the signal generator output frequency until lock occurs.
6. Measure the upper capture frequency (F_{CU}) as follows: Set the signal generator output frequency to 180 kHz. Slowly decrease the signal generator output frequency until lock occurs.
7. Calculate the loop-tracking bandwidth using the following formula.

$$B_t = \frac{R_0}{R_1}$$

where B_t = loop-tracking bandwidth

8. Measure the lower lock limit (F_{LL}) as follows: Set the signal generator output frequency to F_o (i.e., lock occurs). Slowly decrease the signal generator output frequency until the PLL loses lock.
9. Measure the upper lock limit (F_{LU}) as follows: Set the signal generator output frequency to F_o. Slowly increase the signal generator output frequency until the PLL loses lock.
10. Determine the loop-tracking bandwidth using the following formula.

$$B_t = \frac{\Delta F}{F_o}$$

where B_t = loop-tracking bandwidth
$\Delta F = F_{LU} - F_{LL}$
F_o = VCO free-running frequency

11. Replace R_1 with a 100 k-ohm resistor and calculate the loop-tracking bandwidth using the following formula.

$$B_t = \frac{R_0}{R_1}$$

where B_t = loop-tracking bandwidth
R_0 = timing resistor
R_1 = feedback resistor

12. Repeat steps 5 through 10 and compare the measured loop-tracking bandwidth to the value calculated in step 10.

13. Describe the relationship between the resistance of R_1 and the tracking bandwidth.

SECTION D   Phase-Comparator and VCO Conversion Gains

In this section the conversion gains for the PLL's internal phase comparator and VCO are examined. The VCO conversion gain (K_0) is the change in VCO output frequency per unit of dc voltage change at the VCO input (i.e., $K_0 = \Delta F / \Delta V$). The phase-comparator conversion gain (K_θ) is the change in phase-comparator output voltage per unit of phase difference (error) at the phase-comparator input (i.e., $K_\theta = \Delta V / \Delta \theta$). When the external input frequency is equal to the VCO free-running frequency (i.e., $F_o = F_i$), there is a 90° phase difference between F_o and F_i. This 90° phase difference is called the bias or offset phase. As F_o varies above and below F_i, the phase difference between the two signals varies proportionately. The shift in phase from the 90° offset is the phase error (θ_e). θ_e produces a dc voltage (V_d) at the phase-comparator output. V_d is coupled back to the VCO input, where it shifts the VCO frequency until F_o equals F_i. The schematic diagram of the PLL circuit used in this section is the same as the circuit used in Sections A, B, and C and shown in Figure 18-2.

Procedure

1. Construct the PLL circuit shown in Figure 18-2 with R_1 = 10 k ohms and C_0 = 0.001 μF.

2. Adjust the amplitude of the signal generator output voltage to 4 V p–p and set its frequency to the VCO free-running frequency.

3. Measure the phase difference between F_o and F_i with a dual trace oscilloscope.

4. Sketch the waveforms for F_o and F_i.

5. Slowly vary the signal generator output frequency above and below F_o while observing the relative phase error θ_e between F_o and F_i with a dual trace oscilloscope.

6. Describe the relationship between F_i, F_o, and θ_e.

7. Adjust the signal generator output frequency until $F_i = F_o$.

8. Calculate the VCO conversion gain using the following formula.

$$K_0 = \frac{-1}{V_r C R}$$

where K_0 = VCO conversion gain (Hz/V)
V_r = PLL reference voltage
$C = C_0$
$R = R_0$

9. The maximum voltage swing at the phase comparator output is equal to $\pm V_r$, and the maximum phase error is equal to $\pi/2$ radians. Therefore, the maximum phase comparator conversion gain can be calculated using the following formula.

$$K_\theta = \frac{-2V_r}{\pi}$$

where K_θ = maximum phase comparator conversion gain (V/rad)
V_r = PLL reference voltage

10. Measure the VCO and phase comparator conversion gains as follows: Measure the phase comparator output voltage (V_d) on pin 10 when $F_i = F_o$. Slowly increase F_i until the phase error is $-90°$ (i.e., a net phase difference of 180°). Measure the VCO output frequency on pin 5 and phase comparator output voltage and determine the VCO and phase comparator conversion gains using the following formulas.

$$K_0 = \frac{\Delta F}{\Delta V}$$

where K_0 = VCO conversion gain (Hz/V)
ΔF = change in F_o
ΔV = change in V_d

$$K_\theta = \frac{\Delta V}{\Delta \theta}$$

where K_θ = phase comparator conversion gain (V/rad)
ΔV = change in V_d
$\Delta \theta$ = change in phase ($\pi/2$ rad)

11. Repeat step 10, but decrease F_i until the phase error is $+90°$ (i.e., a net phase difference of 0°).

12. Compare the conversion gains measured in steps 10 and 11 to those calculated in steps 8 and 9.

13. Construct a graph of the VCO output-frequency-versus-dc-input voltage characteristics.

14. Construct a graph of the phase comparator output-voltage-versus-phase-error characteristics.

SECTION E   Summary

Write a brief summary of the concepts presented in this experiment on phase-locked loops. Include the following items:

1. The operation of a voltage-controlled oscillator.
2. The operation of a phase comparator.
3. The process of loop acquisition.
4. The concepts of loop lock, pull-in range, capture range, lock range, and hold-in range.
5. VCO conversion gain.
6. Phase-comparator conversion gain.

EXPERIMENT 18   Answer Sheet

NAME: _____ CLASS: _____ DATE: _____

SECTION A

2. F_o (calculated) = _____

3. F_o (measured) = _____

4. % error = _____

5. F_o (calculated) = _____

   F_o (measured) = _____

   % error = _____

6. F_o (calculated) = _____

   F_o (measured) = _____

   % error = _____

7. _____

   _____

   _____

SECTION B

2. F_o (measured) = _____

5. V_r (calculated) = _____

6. V_r (measured) = _____

7. V_d = _____

9. _____

   _____

   _____

SECTION C

2. F_o (measured) = _____

5. F_{CL} = _____

6. F_{CU} = _____

7. B_t (calculated) = _____

8. F_{LL} = _____

9. F_{LU} = _____

10. B_t (measured) = _____

11. B_t (calculated) = _____

12. F_{CL} = _____

   F_{CU} = _____

   B_t (calculated) = _____

   F_{LL} = _____

   F_{LU} = _____

   B_t (measured) = _____

13. _____

_____

_____

SECTION D

3. θ_e = _____

4.

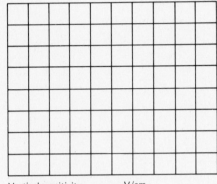

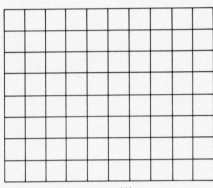

Vertical sensitivity _____ V/cm

Time base _____ sec/cm

Vertical sensitivity _____ V/cm

Time base _____ sec/cm

6. _____

_____

_____

8. K_0 (calculated) = _____   9. K_θ (calculated) = _____

10. K_0 (measured) = _____

K_θ (measured) = _____

11. K_0 (measured) = _____

K_θ (measured) = _____

12. _____

_____

_____

13.

DC input voltage

VCO output frequency

14.

Output voltage

θ

PHASE-LOCKED LOOP FREQUENCY SYNTHESIZERS

REFERENCE TEXT: *Fundamentals of Electronic Communications Systems.*

1. Chapter 5, Phase-Locked Loop.
2. Chapter 5, Indirect Frequency Synthesizers.
3. Chapter 5, Phase-Locked Loop Frequency Synthesizer.

OBJECTIVES

1. To observe the operation of a linear integrated-circuit phase-locked loop.
2. To observe the operation of an indirect frequency synthesizer.
3. To observe the operation of a linear integrated-circuit frequency synthesizer.

INTRODUCTION

A frequency synthesizer is a circuit that can produce many output frequencies through the addition, subtraction, multiplication, and division of a smaller number of fixed frequencies. Essentially, there are two methods of frequency synthesis: direct and indirect. With direct frequency synthesis, multiple output frequencies are generated from two or more crystal-controlled oscillators or by dividing or multiplying the output frequency from a single crystal oscillator. With indirect frequency synthesis, a feedback-controlled divider/multiplier (such as a phase-locked loop) is used to generate multiple output frequencies. In this experiment the operation of a simple single-loop phase-locked-loop (PLL) frequency synthesizer is examined. The XR-2212 precision phase-locked loop is used for the feedback control circuit, and a CD4027B dual JK flip-flop is used for the divider circuit. The block diagrams for the XR-2212 and CD4027B are shown in Figure 19-1.

MATERIALS REQUIRED

Equipment:

1 — protoboard
1 — dc power supply (12 V dc)

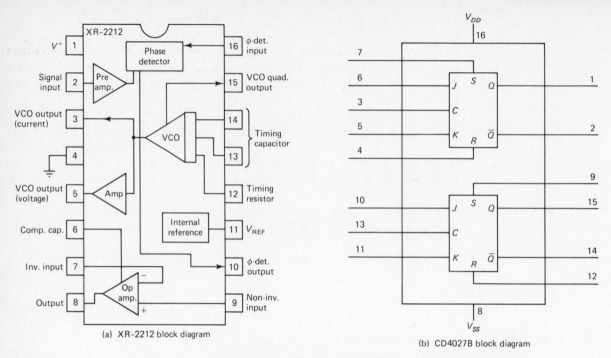

FIGURE 19-1 Functional block diagrams, (a) XR-2212, (b) CD4027B

1 — medium-frequency signal generator (1 MHz)

1 — standard oscilloscope (10 MHz)

1 — assortment of test leads and hookup wire

Parts List:

1 — XR-2212 monolithic phase-locked loop

1 — CD4027B dual JK flip-flop

1 — 1 k-ohm resistor

2 — 10 k-ohm resistors

1 — 22 k-ohm resistor

1 — 100 k-ohm resistor

1 — 0.001-μF capacitor

1 — 0.05-μF capacitor

2 — 0.1-μF capacitors

SECTION A   Phase-Locked Loop Operation

In this section the basic operation of the XR-2212 linear integrated-circuit phase-locked loop (PLL) is examined. A PLL is a closed-loop feedback control system in which the feedback signal is a frequency rather than simply a dc voltage. The XR-2212 is an ultrastable monolithic phase-locked-loop system especially designed for data-communications and control-system applications. The block diagram for the XR-2212 is shown in Figure 19-1a. The XR-2212 comprises four primary sections: an input preamplifier, a phase detector, a voltage-controlled oscillator (VCO), and a high-gain differential amplifier. A VCO is a free-running oscillator with a stable frequency of oscillation (F_o) that is dependent on a timing capacitance, a timing resistance, and an external control voltage.

The output from a VCO is a frequency, and the input is a bias or control voltage. In essence, a VCO performs voltage-to-frequency conversion. A phase detector (or phase comparator as it is sometimes called) is a balanced mixer with two inputs: an external input frequency (F_i) and the VCO output frequency (F_o). Once frequency lock has occurred, the phase comparator output is a dc voltage that is proportional to the phase difference (θ_e) between F_o and F_i. The basic operation of a phase-locked loop is quite simple. The external input signal (F_i) is compared to the VCO output signal (F_o) in the phase comparator. The phase comparator output voltage is applied to the VCO input, where it deviates (shifts) the frequency of the VCO until $F_o = F_i$. This process is called loop acquisition. Once $F_o = F_i$, the loop is said to be locked and the F_o will follow changes in F_i. The block diagram for the PLL circuit used in this section is shown in Figure 19-2.

Procedure

1. Construct the PLL circuit shown in Figure 19-2.
2. Calculate the VCO free-running frequency using the following formula.

$$F_o = \frac{1}{RC}$$

where F_o = free-running frequency
$\quad\quad R = R_0$
$\quad\quad C = C_0$

3. Disconnect one end of resistor R_1 and measure the VCO free-running frequency on pin 5.
4. Calculate the percentage error between the calculated and actual free-running frequencies using the following formula.

$$\% \text{ error} = \frac{\text{measured value} - \text{calculated value}}{\text{calculated value}} \times 100$$

5. Reconnect R_1 and set the amplitude of the signal generator output voltage to 4 V p–p and adjust its frequency to 30 kHz.
6. Slowly increase the signal generator output frequency while observing both F_o and F_i. Note that as F_i approaches F_o, the PLL locks onto the signal generator output

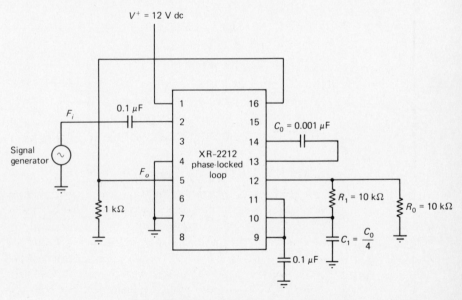

FIGURE 19-2   XR-2212 Phase-locked loop

frequency and follows (tracks) it. However, if F_i is increased significantly above F_o, the PLL loses lock. After the PLL loses lock, slowly decrease the F_i. Note that as F_i approaches F_o, the PLL locks onto F_i and follows it. Again, if F_i is decreased significantly below F_o, the PLL loses lock.

7. Repeat step 6, but observe the phase-comparator output voltage (V_d) on pin 10.
8. Describe the relationship between F_i, F_o, and V_d.

SECTION B PLL Frequency Synthesizer

In this section the operation of an indirect PLL frequency synthesizer is examined. In essence, an indirect PLL frequency synthesizer performs times-N frequency multiplication, where N is any integer value greater than unity. The PLL frequency synthesizer circuit used in this section is shown in Figure 19-3. The VCO output signal (F_o) is applied to the input of a JK flip-flop, which is operating in the T mode and simply divides F_o by two. The output from the JK flip-flop ($F_c = F_o/2$) is applied to one input of the phase detector, where it is compared to F_i. The phase comparator produces an output voltage that is proportional to the difference in phase (θ_e) between F_c and F_i. Therefore, once lock has occurred, $F_o = 2 F_i$.

Procedure

1. Construct the indirect PLL frequency synthesizer circuit shown in Figure 19-3.
2. Calculate the VCO free-running frequency using the following formula.

$$F_o = \frac{1}{RC}$$

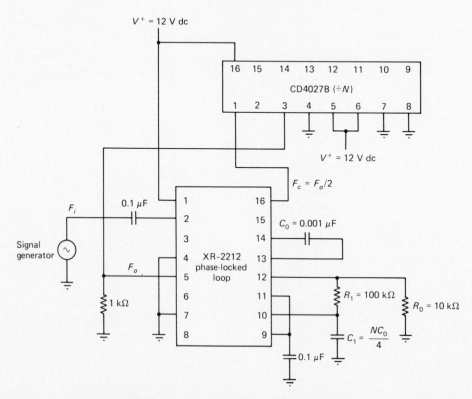

FIGURE 19-3 Phase-locked loop frequency synthesizer

where F_o = free-running frequency
$R = R_0$
$C = C_0$

3. Disconnect one end of resistor R_1 and measure the VCO free-running frequency on pin 5.

4. Reconnect R_1, and set the amplitude of the signal generator output voltage to 4 V p–p and adjust its frequency to 40 kHz.

5. Slowly increase the signal generator output frequency (F_i) while observing the phase-comparator input signals on pins 16 and 2 (F_i and F_c). Once lock has occurred, measure the frequencies of F_i and F_c.

6. Measure the VCO output frequency (F_o) on pin 5 of the XR-2212.

7. Describe the relationship between F_i, F_c, and F_o.

8. Slowly vary the signal generator output frequency (F_i) first above and then below F_o while observing F_c and F_o.

9. Describe what effect varying F_i had on F_c and F_o.

10. Change the divide-by-2 circuit shown in Figure 19-3 to a divide-by-4 circuit as shown in Figure 19-4.

11. Set the amplitude of the signal generator output voltage to 4 V p–p and adjust its frequency to 5 kHz.

12. Repeat steps 5 through 9.

13. Design a divide-by-3 frequency synthesizer using the CD4027B dual JK flip-flop and the XR-2212 PLL.

14. Construct the circuit designed in step 13 and verify its operation.

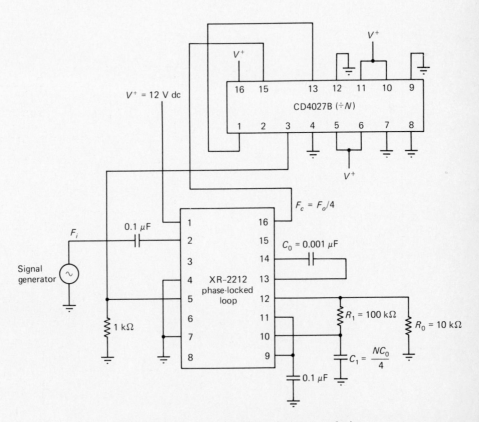

FIGURE 19-4 Phase-locked loop frequency synthesizer

SECTION C Summary

Write a brief summary of the concepts presented in this experiment on indirect PLL frequency synthesizers. Include the following items:

1. The basic operation of a phase-locked loop.
2. The concept of loop acquisition.
3. The concept of indirect frequency synthesis.
4. The concept of frequency division and frequency multiplication.

EXPERIMENT 19 Answer Sheet

NAME: _____ CLASS: _____ DATE: _____

SECTION A

2. F_o (calculated) = _____   3. F_o (measured) = _____

4. % error = _____

8. _____

_____

_____

SECTION B

2. F_o (calculated) = _____   3. F_o (measured) = _____

5. F_c (measured) = _____   6. F_o (measured) = _____

   F_i (measured) = _____

7. _____

_____

_____

9. _____

_____

_____

12. F_i (measured) = _____

    F_c (measured) = _____

    F_o (measured) = _____

_____

_____

_____

_____

_____

_____

Experiment 20

LINEAR INTEGRATED-CIRCUIT BALANCED MODULATOR

REFERENCE TEXT: *Fundamentals of Electronic Communications Systems.*

1. Chapter 6, AM Double Sideband Suppressed Carrier (AM DSBSC).
2. Chapter 6, Linear Integrated-Circuit (LIC) Balanced Modulator.

OBJECTIVES

1. To observe the operation of a linear integrated-circuit balanced modulator.
2. To observe the operation of a LIC DSBSC modulator.
3. To observe the operation of a LIC DSBFC modulator.
4. To observe the operation of a LIC frequency doubler.

INTRODUCTION

Integrated-circuit balanced modulators are ideally suited for applications that require balanced operation. In addition, integrated-circuit balanced modulators offer low power consumption, circuit miniaturization, and simplicity of design. In this exercise the operation of the LM1496/1596 silicon monolithic linear integrated circuit is examined. The LM1496/1596 is a double balanced modulator-demodulator that produces an output voltage that is proportional to the product of an input (signal) voltage and a switching (carrier) signal. The LM1496/1596 features excellent carrier suppression, adjustable gain and signal handling, fully balanced inputs and outputs, low offset and drift, and wide frequency response. The block diagram for the test circuit used in this experiment is shown in Figure 20-1. The LM1496/1596 is used both as an AM DSBSC (product) modulator

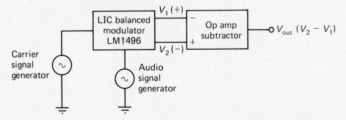

FIGURE 20-1 Linear integrated-circuit balanced modulator test circuit

149

and an AM DSBFC modulator. The modulators have two input signals: a carrier and a modulating (audio) signal.

MATERIALS REQUIRED

Equipment:

1 — protoboard
1 — dual dc power supply (-8 V dc and $+12$ V dc)
1 — audio signal generator
1 — medium-frequency signal generator (1 MHz)
1 — standard oscilloscope (10 MHz)
1 — assortment of test leads and hookup wire

Parts List:

1 — LM1496/1596 LIC balanced modulator
1 — LIC operational amplifier (741 or equivalent)
3 — 51-ohm resistors
2 — 100-ohm resistors
4 — 1 k-ohm resistors
1 — 1.2 k-ohm resistor
2 — 3.9 k-ohm resistors
1 — 6.8 k-ohm resistor
1 — 10 k-ohm resistor
1 — 5 k-ohm variable resistor
1 — 10 k-ohm variable resistor
4 — equal-value resistors (1 to 10 k ohms)
1 — 0.001-μF capacitor
1 — 0.1-μF capacitor

SECTION A   Double Sideband Suppressed Carrier Modulator

In this section the operation of a linear integrated-circuit, double sideband suppressed carrier (DSBSC) modulator is examined. The LM1496 is a balanced modulator, which is a product modulator. With a product modulator the output waveform is the product of the two input signals (the carrier signal and the modulating signal). Therefore, the output waveform contains the upper and lower side frequencies only; the carrier and modulating signals are suppressed. The schematic diagram for the DSBSC modulator used in this section is shown in Figure 20-2. The LM1496 and its associated circuitry is a DSBSC modulator. The LM1496 provides between 40 and 60 dB of carrier suppression. Pin 6 (V_1) and pin 12 (V_2) are single-ended outputs containing the sum and difference frequencies and a small component of the carrier. V_1 and V_2 are 180° out of phase ($V_1 = -V_2$). The op amp circuit is a unity-gain subtractor. V_1 is inverted and added to V_2. Therefore, V_1 is effectively subtracted from V_2, and $V_{out} = V_2 - V_1$; or $V_2 - (-V_2) = 2 V_2$. R_1 is a balance control which is used to suppress the carrier. Maximum carrier suppression is achieved with R_1 set to midrange. Minimum carrier suppression is achieved when R_1 is set to either its maximum clockwise or its maximum counterclockwise position. Rotating R_1 from its maximum clockwise to its maximum counterclockwise position simply reverses the phase of V_1 and V_2. Consequently, for balanced operation (R_1 set to midrange), the op

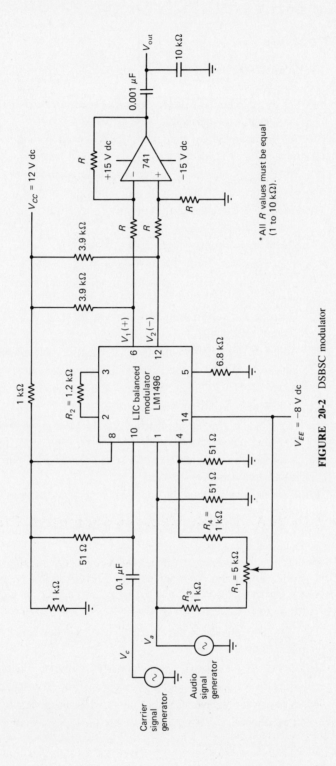

FIGURE 20-2 DSBSC modulator

151

amp output (V_{out}) contains the same frequency components as V_1 and V_2 but is equal to approximately twice the amplitude of either. Variable resistor R_2 sets the gain of the LM1496.

Procedure

1. Construct the double sideband suppressed carrier modulator circuit shown in Figure 20-2.
2. Set the amplitude of the carrier signal generator output voltage to 1 V p–p and a frequency $F_c = 40$ kHz.
3. Set the amplitude of the audio signal generator output voltage to 0 V and a frequency $F_a = 1$ kHz.
4. Set R_1 to its maximum counterclockwise position.
5. Sketch the waveforms for V_1, V_2 and V_{out}.
6. Describe the relationship between V_1, V_2, and V_{out} in terms of amplitude, frequency, and phase.
7. Rotate R_1 to its maximum clockwise position.
8. Sketch the waveforms for V_1, V_2, and V_{out}.
9. Compare the waveforms sketched in steps 5 and 8 in terms of frequency content and phase.
10. Describe the relationship between V_1, V_2, and V_{out} in terms of amplitude, frequency, and phase.
11. Adjust R_1 for a minimum signal at V_{out} (i.e., maximum carrier suppression).
12. Set the amplitude of the audio signal generator output voltage to 1 V p–p.
13. Adjust R_1 for a DSBFC envelope with minimum distortion.
14. Sketch the waveforms for V_a and V_{out}.
15. Describe the waveforms sketched in step 14 in terms of amplitude, frequency content, and repetition rate.
16. Vary the amplitude of the carrier signal generator output voltage and describe what effect varying it has on V_{out}.
17. Vary the amplitude of the audio signal generator output voltage and describe what effect varying it has on V_{out}.
18. Vary the frequency of the audio signal generator output and describe what effect varying it has on V_{out}.
19. Reduce the frequency of the carrier signal generator output to 10 kHz and sketch the waveform at V_{out} (set the frequency of the audio signal generator to 1 kHz).
20. Describe the waveform sketched in step 19 in terms of frequency content and repetition rate.
21. Set the amplitude of the carrier signal generator output voltage to 1 V p–p and a frequency $F_c = 40$ kHz.
22. Adjust R_1 for a DSBFC envelope with minimum distortion.
23. Replace R_2 with a 5 k-ohm variable resistor (rheostat).
24. Vary R_2 throughout its entire range while observing V_{out}.
25. Describe what effect varying R_2 has on V_{out}.

SECTION B   Double Sideband Full Carrier Modulator

In this section the operation of a linear integrated-circuit, double sideband full carrier (DSBFC) AM modulator is examined. The suppressed carrier modulator examined in

Section A and shown in Figure 20-2 is easily converted to a double sideband full carrier modulator by simply extending the balance range. This is accomplished by increasing the ratio of R_1 to R_3 and of R_1 to R_4 (i.e., reducing the resistance of R_2 and of R_4). This allows the LM1496 to be operated sufficiently out of balance to produce a significant carrier component at its outputs.

Procedure

1. Replace R_3 and R_4 in Figure 20-2 with 100-ohm resistors.
2. Replace variable resistor R_2 with a 1.2 k-ohm resistor.
3. Set the amplitude of the carrier signal generator output voltage to 1 V p–p and a frequency $F_c = 100$ kHz.
4. Set the amplitude of the audio signal generator output voltage to 100 mV p–p and a frequency $F_a = 1$ kHz.
5. Adjust R_1 until an AM DSBFC envelope is observed at V_{out} with minimum distortion.
6. Adjust the amplitude of the audio signal generator output voltage until a 100% AM modulated envelope is observed at V_{out}.
7. Sketch the waveform for v_a and V_{out}.
8. Describe the waveforms sketched in step 7 in terms of amplitude and frequency content.
9. Vary the frequency and amplitude of the audio signal generator output voltage while observing V_{out}.
10. Describe what effect varying the amplitude and frequency of the audio signal generator has on the output waveform.

SECTION C   Linear Integrated-Circuit Frequency Doubler

In this section the operation of a linear integrated-circuit frequency doubler is examined. When operated in the balanced mode, a linear integrated-circuit balanced modulator suppresses the two input signals and produces an output waveform that is the product of the two input frequencies (i.e., the sum and difference frequencies). If the same frequency is applied to both inputs of a balanced modulator, the output contains the sum frequency (i.e., a signal with a frequency equal to twice the input frequency) and a difference frequency (i.e., 0 Hz). The schematic diagram for the LIC frequency doubler circuit used in this section is shown in Figure 20-3. The signal generator output frequency is applied to both inputs of the balanced modulator, producing sum and difference frequencies at its output. The sum frequency is equal to twice the signal generator frequency, and the difference frequency is 0 Hz.

Procedure

1. Construct the LIC frequency doubler circuit shown in Figure 20-3.
2. Set the amplitude of the signal generator output voltage to 0.25 V p–p and a frequency $F_{in} = 10$ kHz.
3. Adjust R_1 and R_2 for an output signal on pin 6 of the balanced modulator with minimum distortion and a frequency $F_{out} = 2 F_{in}$.
4. Sketch the waveform observed in step 3.
5. Describe the waveform sketched in step 4 in terms of frequency, phase, and amplitude.

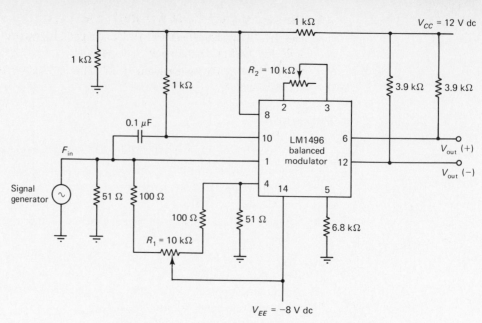

FIGURE 20-3 Linear integrated-circuit frequency doubler

6. Vary R_1 over its entire range and describe what effect varying it has on the waveform observed on pin 6.

7. Repeat step 3.

8. Sketch the waveform observed on pin 12 of the balanced modulator.

9. Describe the waveform sketched in step 8 in terms of frequency, phase, and amplitude.

10. Vary R_1 over its entire range and describe what effect varying it has on the waveform on pin 12.

11. Compare the waveforms sketched in steps 4 and 8 in terms of frequency, phase, and amplitude.

12. Vary R_2 over its entire range and describe what effect varying it has on the waveform observed on pins 6 and 12 of the balanced modulator.

13. Vary the signal generator output frequency and describe what effect varying it has on the waveform observed on pins 6 and 12 of the balanced modulator.

14. Vary the amplitude of the signal generator output voltage and describe what effect varying it has on the waveforms observed on pins 6 and 12 of the balanced modulator.

SECTION D Summary

Write a brief summary of the concepts presented in this experiment on linear integrated-circuit balanced modulators. Include the following items:

1. The frequency content of a DSBSC waveform.

2. The relationship between the modulating signal, the carrier, and the output envelope with DSBSC modulation.

3. The concept of carrier suppression.

4. The concept of balanced operation.

5. The frequency content of a DSBFC waveform.

6. The relationship between the modulating signal, the carrier, and the output envelope with AM DSBFC modulation.

7. The concept of frequency multiplication using a balanced modulator.

EXPERIMENT 20   Answer Sheet

NAME: _____ CLASS: _____ DATE: _____

SECTION A

5.

Vertical sensitivity _____ V/cm

Time base _____ sec/cm

Vertical sensitivity _____ V/cm

Time base _____ sec/cm

Vertical sensitivity _____ V/cm

Time base _____ sec/cm

6. _____

_____

_____

8.

Vertical sensitivity _____ V/cm

Time base _____ sec/cm

Vertical sensitivity _____ V/cm

Time base _____ sec/cm

Vertical sensitivity _____ V/cm

Time base _____ sec/cm

9. _____
_____
_____

10. _____
_____
_____

14.

Vertical sensitivity _____ V/cm

Time base _____ sec/cm

Vertical sensitivity _____ V/cm

Time base _____ sec/cm

15. _____
_____
_____

16. _____
_____
_____

17. _____
_____
_____

18. _____
_____
_____

19.

Vertical sensitivity _____ V/cm

Time base _____ sec/cm

Vertical sensitivity _____ V/cm

Time base _____ sec/cm

20. _____

_____

_____

25. _____

_____

_____

SECTION B

7.

Vertical sensitivity _____ V/cm

Time base _____ sec/cm

Vertical sensitivity _____ V/cm

Time base _____ sec/cm

8. _____

_____

_____

10. _____

_____

_____

SECTION C

4.

Vertical sensitivity _____ V/cm

Time base _____ sec/cm

5. _____

_____

_____

8.

Vertical sensitivity _____ V/cm

Time base _____ sec/cm

9. _____

_____

_____

10. _____

_____

_____

11. _____

_____

_____

12. _____

_____

_____

13. _____

_____

_____

14. _____

_____

_____

LINEAR INTEGRATED-CIRCUIT FREQUENCY MODULATOR

REFERENCE TEXT: *Fundamentals of Electronic Communications Systems.*
Chapter 7, Linear Integrated-Circuit Direct FM Generator.

OBJECTIVES

1. To observe the operation of a direct FM modulator.
2. To observe the operation of a linear integrated-circuit FM generator.

INTRODUCTION

Linear integrated-circuit (LIC) FM modulators generate a high quality output waveform that is stable, accurate, and directly proportional to the input modulating signal. In this experiment the operation of the XR-2206 function generator as a direct FM modulator is examined. The block diagram for the linear integrated-circuit direct FM modulator used in this experiment is shown in Figure 21-1. The XR-2206 function generator comprises four functional blocks: a voltage-controlled oscillator (VCO), an analog multiplier and sine shaper, a unity-gain buffer amplifier, and a set of current switches. The VCO free-running frequency is the carrier rest frequency and is determined by an external timing resistor and an external timing capacitor. The input modulating signal deviates the center frequency, which produces an FM output wave.

MATERIALS REQUIRED

Equipment:

1 — protoboard
1 — dual dc power supply (+ 12 V dc and − 5 to + 5 V dc)
1 — audio signal generator
1 — standard oscilloscope (10 MHz)
1 — assortment of test leads and hookup wire

Note: If the audio signal generator is equipped with a variable dc offset, the − 5 to + 5 V dc supply is not needed.

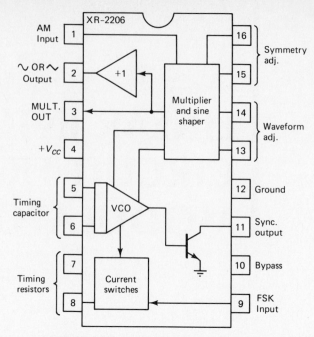

FIGURE 21-1   XR-2206 Functional block diagram

Parts List:

1 — XR-2206 monolithic function generator

3 — 4.7 k-ohm resistors

2 — 10 k-ohm resistors

1 — 47 k-ohm resistor

1 — 1 k-ohm variable resistor

2 — 0.001-µF capacitors

2 — 1-µF capacitors

1 — 10-µF capacitor

SECTION A   Deviation Sensitivity (Voltage-to-Frequency Conversion)

In this section the deviation sensitivity of a linear integrated-circuit direct FM modulator is examined. The XR-2206 function generator is used for the modulator circuit. In essence, the deviation sensitivity of a voltage-controlled oscillator is its transfer function (i.e., its output-versus-input characteristics). The input to a voltage-controlled oscillator is a bias or control voltage, and the output is a frequency. Therefore, the deviation sensitivity for a VCO FM modulator is the change in output frequency divided by the change in input voltage (i.e., k_1 = Hz/V). In essence, the deviation sensitivity of a VCO is its voltage-to-frequency conversion ratio. In actuality, the frequency of oscillation for the XR-2206 function generator is controlled by its input current. However, the input current can be controlled with an extrernal bias voltage. The frequency of oscillation for the XR-2206 varies inversely with the polarity of the bias voltage (i.e., a positive bias voltage decreases its frequency and a negative bias voltage increases its frequency). The schematic diagram for the linear integrated-circuit direct FM modulator used in this section is shown in Figure 21-2. The VCO free-running frequency is the carrier rest frequency.

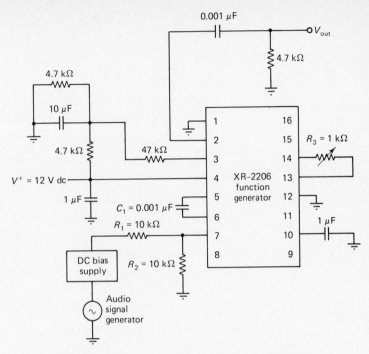

FIGURE 21-2 Linear integrated-circuit direct FM modulator

Procedure

1. Construct the linear integrated-circuit direct FM modulator circuit shown in Figure 21-2.
2. Set the bias voltage (V_C) to 0 V dc.
3. Vary R_3 until a sine wave with minimum distortion is observed at V_{out}.
4. Calculate the carrier rest frequency using the following formula.

$$F_c = \frac{1}{RC}$$

where F_c = carrier rest frequency

$$R = \frac{R_1 \times R_2}{R_1 + R_2}$$
$$C = C_1$$

5. Measure the carrier rest frequency at V_{out}.
6. Determine the percentage error between the calculated and actual carrier rest frequencies using the following formula.

$$\% \text{ error} = \frac{\text{measured value} - \text{calculated value}}{\text{calculated value}} \times 100$$

7. Vary the bias voltage first in the positive and then in the negative direction while observing V_{out}.
8. Describe the relationship between the magnitude and polarity of the bias voltage and the VCO output frequency.
9. Measure the VCO output frequency (F_o) for a bias voltage $V_C = +2$ V dc.

10. Determine the deviation sensitivity of the VCO using the following formula.

$$k_1 = \frac{\Delta F}{\Delta V} = \frac{F_c - F_o}{0 - 2}$$

where k_1 = deviation sensitivity (Hz/V)
   F_c = carrier rest frequency (V_C = 0 V dc)
   F_o = output frequency (V_C = +2 V dc)

11. Calculate the VCO output frequency for a bias voltage V_C = -2 V dc using the following formula.

$$F_o = F_c + \Delta F$$

where F_o = frequency of oscillation
   F_c = carrier rest frequency
   $\Delta F = V_C \times k_1$

12. Measure the actual output frequency for a bias voltage V_C = -2 V dc.
13. Determine the percentage error between the calculated and actual output frequencies for a bias voltage V_C = -2 V dc.

SECTION B   Frequency Deviation

In this section the frequency deviation for the linear integrated-circuit direct FM modulator is examined. The modulator circuit used in this section is the same circuit used in Section A and shown in Figure 21-2. Frequency deviation (ΔF) is the change that occurs in the carrier frequency when it is acted on by a modulating signal. In a direct FM modulator, the frequency deviation is proportional to the amplitude of the modulating signal voltage (V_a), and the rate at which the frequency change occurs is equal to the modulating signal frequency (F_a). Therefore, frequency deviation is a function of the deviation sensitivity of the modulator and the amplitude of the modulating signal (i.e., $\Delta F = V_a k_1$). Figure 21-3 illustrates the concept of frequency deviation. As the carrier frequency increases, the period of the waveform decreases and vice versa. This relationship is shown in Figure 21-3a. When the rate of deviation is increased (i.e., F_a increases), the modulated carrier resembles the waveform shown in Figure 21-3b. This waveform graphically illustrates the deviation in the VCO output frequency. The "fat" portion of the waveform corresponds to the peak-to-peak change in the period of the carrier. The minimum period (T_1) corresponds to the maximum output frequency, and the maximum period (T_2) corresponds to the minimum output frequency. Therefore, the peak-to-peak frequency deviation (sometimes called carrier swing) is determined by simply measuring the difference between the maximum and minimum output frequencies (i.e., $1/T_1 - 1/T_2$). The peak frequency deviation (ΔF) is found by simply dividing the carrier swing by two.

Procedure

1. Construct the linear integrated-circuit direct FM modulator circuit shown in Figure 21-3.
2. Set the dc bias supply (V_C) to 0 V and adjust the oscilloscope time base until two or three cycles of the carrier frequency are observed at V_{out}.
3. Adjust the trigger slope and level controls and the horizontal position control on the oscilloscope until the waveform begins at the left most gradicule as shown in Figure 21-3a.
4. Measure the frequency of the waveform observed at V_{out}.
5. Vary the dc bias supply first in the positive and then in the negative direction while observing V_{out}.

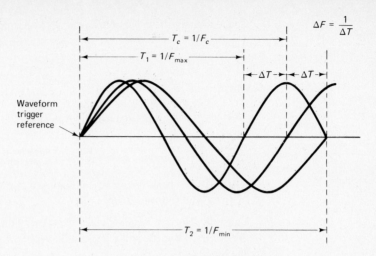

(a) Frequency deviation

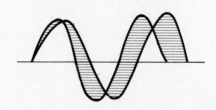

(b) Frequency modulation

FIGURE 21-3 FM Generation. (a) Frequency deviation. (b) Frequency modulation.

6. Describe what effect varying the dc bias voltage has on the waveform observed at V_{out}.

7. Adjust the dc bias supply (V_C) to 0 V dc and set the amplitude of the audio signal generator output voltage to 2 V p–p and a frequency $F_a = 1$ Hz.

8. Describe the waveform observed at V_{out}.

9. Slowly increase the audio signal generator output frequency while observing V_{out}.

10. Describe the waveform observed in step 9.

11. Increase the audio signal generator output frequency to 100 Hz. The oscilloscope triggers each sweep at the same point on the display. However, since each cycle has a slightly different frequency, the waveform appears blurred.

12. Measure the maximum period of the output waveform and determine the minimum VCO output frequency using the following formula.

$$F_{min} = \frac{1}{T_{max}}$$

where F_{min} = minimum output frequency
T_{max} = maximum output period

13. Measure the minimum period of the output waveform and determine the maximum VCO output frequency using the following formula.

$$F_{max} = \frac{1}{T_{min}}$$

where F_{max} = maximum output frequency
T_{min} = minimum output period

14. Calculate the carrier swing and peak frequency deviation using the following formulas.

$$CS = F_{max} - F_{min}$$

where CS = carrier swing
F_{max} = maximum output frequency
F_{min} = minimum output frequency

$$\Delta F = \frac{CS}{2}$$

where ΔF = peak frequency deviation
CS = carrier swing

15. Calculate the peak frequency deviation using the deviation sensitivity determined in Section A, step 10 and a peak audio voltage V_a = 1 V p.

16. Compare the peak frequency deviation calculated in step 15 to the value calculated in step 14.

SECTION C   SUMMARY

Write a brief summary of the concepts presented in this experiment on linear integrated-circuit frequency modulators. Include the following items:

1. The concept of deviation sensitivity and voltage-to-frequency conversion.
2. The concept of frequency deviation.
3. The operation of a linear integrated-circuit VCO.
4. The operation of a linear integrated-circuit frequency modulator.

EXPERIMENT 21   Answer Sheet

NAME: _____ CLASS: _____ DATE: _____

SECTION A

4. F_c (calculated) = _____   5. F_c (measured) = _____

6. % error = _____

8. _____

_____

_____

9. F_o (measured) = _____   10. k_1 = _____

11. F_o (calculated) = _____   12. F_o (measured) = _____

13. % error = _____

SECTION B

4. F_{out} (measured) = _____

6. _____

_____

_____

8. _____

_____

10. _____

_____

_____

12. F_{min} = _____   13. F_{max} = _____

14. CS = _____ ΔF = _____   15. ΔF = _____

16. _____

_____

_____

LINEAR INTEGRATED-CIRCUIT FM DEMODULATOR

REFERENCE TEXT: *Fundamentals of Electronic Communications Systems.*

1. Chapter 8, FM Demodulators.
2. Chapter 8, PLL FM Demodulator.

OBJECTIVES

1. To observe the process of FM demodulation.
2. To observe the operation of a linear integrated-circuit FM demodulator.
3. To observe the operation of a PLL FM demodulator.

INTRODUCTION

Since the development of linear integrated circuits, FM demodulation can be accomplished quite simply with a phase locked loop (PLL). A PLL frequency demodulator requires no tuned circuits and automatically compensates for carrier shift due to instability of the transmit oscillator. In this experiment FM demodulation is accomplished with an XR-2212 linear integrated-circuit phase locked loop. The XR-2212 PLL can be used as a linear FM demodulator for both narrow-band and wide-band FM signals. The block diagram for the XR-2212 PLL is shown in Figure 22-1a. The input FM signal is connected to the input of the internal preamplifier (pin 2), and the VCO output (pin 5) is connected directly to the phase-detector input (pin 16). The demodulated signal appears at the phase-detector output (pin 10), which is directly connected to the PLL internal op amp. The op amp is used as a buffer amplifier to provide both additional amplification as well as current drive capability. An XR-2206 function generator is used for the FM modulator. The block diagram for the XR-2206 is shown in Figure 22-1b.

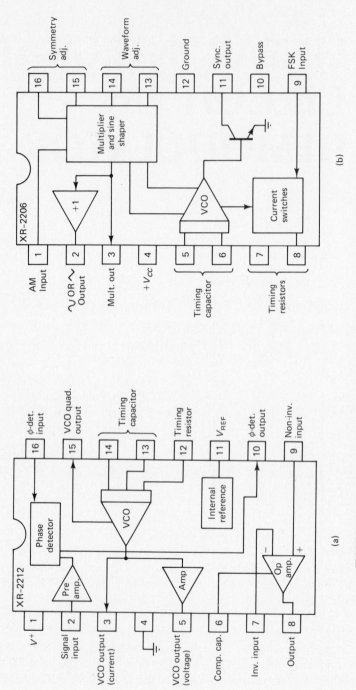

FIGURE 22-1 Functional block diagrams. (a) XR-2212 PLL. (b) XR-2206 Function generator.

MATERIALS REQUIRED

Equipment:

1 — protoboard

1 — dual dc power supply (+12 V dc and −3 to +3 V dc)

1 — medium-frequency signal generator (1 MHz)

1 — standard oscilloscope (10 MHz)

1 — assortment of test leads and hookup wire

Parts List:

1 — XR-2206 monolithic function generator

1 — XR-2212 monolithic phase-locked loop

5 — 4.7 k-ohm resistors

3 — 10 k-ohm resistors

2 — 47 k-ohm resistors

2 — 100 k-ohm resistors

1 — 1 k-ohm variable resistor

1 — 10 k-ohm variable resistor

1 — 30-pF capacitor

5 — 0.001-μF capacitors

3 — 0.1-μF capacitors

2 — 1-μF capacitors

1 — 10-μF capacitor

SECTION A   Frequency-to-Voltage Conversion Gain

In this section the frequency-to-voltage conversion gain for the XR-2212 PLL is examined. In essence, an FM demodulator performs frequency-to-voltage conversion. However, the phase comparator in a PLL produces an output voltage (V_d) that is proportional to the difference in phase between the external input signal (F_i) and the internal VCO signal (F_o). V_d controls both the frequency and the phase of the VCO output signal. In essence, a PLL converts frequency variations to phase variations, then converts the phase variations to voltage changes. If the external input to the PLL is an FM signal (i.e., a deviated carrier), the magnitude, direction, and rate of change of V_d are proportional to the modulating signal (i.e., the PLL demodulates the FM signal). For symmetrical frequency discrimination, the change in V_d for a given increase in F_i should be equal to the change in V_d for a corresponding decrease in F_i. However, the voltage change should be in the opposite direction (i.e., if an increase in F_i produces a more positive output voltage, a decrease in F_i should produce a more negative output voltage). The schematic diagram for the PLL FM demodulator circuit used in this section is shown in Figure 22-2. The output of the phase comparator is connected to the PLL's internal op amp to amplify the demodulated signal.

Procedure

1. Construct the PLL FM demodulator circuit shown in Figure 22-2 (set the resistance of R_x to midrange).
2. Disconnect one end of resistor R_1.

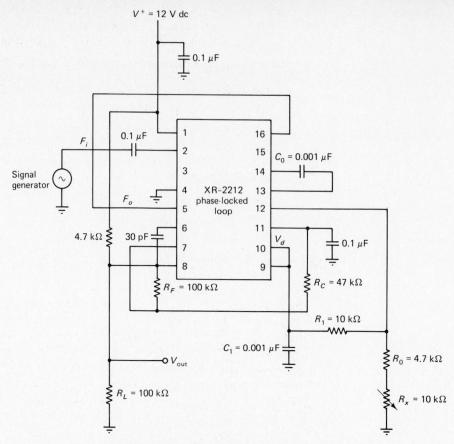

FIGURE 22-2 PLL Frequency-to-voltage conversion

3. Adjust the resistance of R_x for a 100 kHz square wave at F_o (pin 5).

4. Reconnect R_1, and set the amplitude of the signal generator output voltage to 2 V p–p and set its frequency to a 100 kHz sine wave.

5. Verify that frequency lock has occurred by observing F_o and F_i with an oscilloscope.

6. Measure the dc voltage at V_{out}.

7. Increase the signal generator output frequency to 116.7 kHz and measure the dc voltage at V_{out}.

8. Determine the frequency-to-voltage conversion gain for increases in F_i using the following formula.

$$K_+ = \frac{\Delta V}{\Delta F}$$

where K_+ = positive frequency-to-voltage conversion gain
ΔF = 116.7 kHz − 100 kHz = 16.7 kHz
ΔV = the difference between V_{out} measured in step 6 and step 7.

9. Decrease the signal generator output frequency to 83.3 kHz and measure V_{out}.

10. Determine the frequency-to-voltage conversion gain for decreases in F_i using the following formula.

$$K_- = \frac{\Delta V}{\Delta F}$$

where K_- = negative frequency-to-voltage conversion gain
ΔF = 100 kHz − 83.3 kHz = 16.7 kHz
ΔV = the difference between V_{out} measured in step 7 and step 10.

11. The frequency-to-voltage conversion gain for an FM demodulator is generally the average of the positive and negative conversion gains and is calculated using the following formula.

$$K_t = \frac{K_+ - K_-}{2}$$

where K_+ = positive conversion gain
K_- = negative conversion gain
K_t = overall conversion gain

SECTION B   Circuit Gain

In this section the overall circuit gain for an FM transmission system is examined. The circuit gain for an FM transmission system is the ratio of the voltage change at the input to the FM modulator to the corresponding voltage change produced at the output of the FM receiver. In this section a linear integrated-circuit direct FM modulator is used to generate the FM signal, and a linear integrated-circuit PLL is used for the FM demodulator. The schematic diagram for the FM modulator circuit used in this section is shown in Figure 22-3a, and the schematic diagram for the FM demodulator circuit is shown in Figure 22-3b. The modulator is an XR-2206 function generator and the demodulator is an XR-2212 PLL.

Procedure

1. Construct the FM modulator and demodulator circuits shown in Figures 22-3a and 22-3b, respectively.
2. Calculate the modulator carrier rest frequency using the following formula.

$$F_c = \frac{1}{RC}$$

where F_c = carrier rest frequency
$R = R_1$
$C = C_1$

3. Set the dc bias voltage (V_C) and signal generator output voltage to 0 V and adjust R_2 for an undistorted sine wave at the FM output (F_c).
4. Measure the carrier rest frequency (F_c) at the modulator output.
5. Calculate the demodulator VCO free-running frequency (F_o) using the following formula.

$$F_o = \frac{1}{RC}$$

where F_o = VCO free-running frequency
$R = R_o + R_x/2$
$C = C_o$

6. Disconnect one end of R_1 and adjust R_x until the VCO output frequency (F_o) is equal to the carrier rest frequency (F_c) measured in step 4.

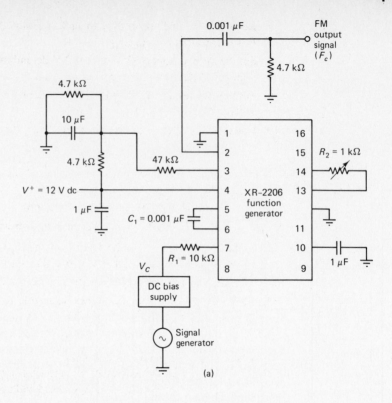

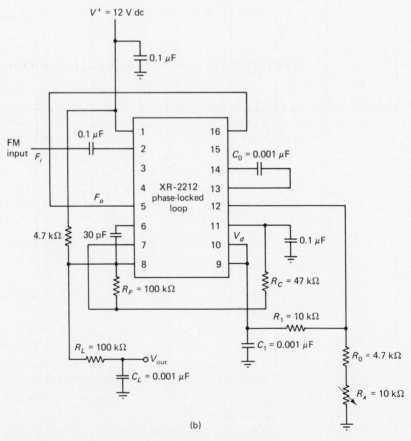

FIGURE 22-3   FM transmission system. (a) FM modulator. (b) FM demodulator.

7. Reconnect R_1 and connect a jumper wire from the FM modulator output (F_c) to the FM demodulator input (F_i).

8. Set the dc bias supply (V_C) to +0.2 V dc and measure the demodualtor dc output voltage (V_{out}).

9. Set the dc bias supply (V_C) to −0.2 V dc and measure the demodulator dc output voltage (V_{out}).

10. Calculate the static circuit gain using the following formula.

$$A_s = \frac{V_{out}}{V_{in}}$$

where A_s = static circuit gain
V_{in} = (+0.2) − (−0.2) = 0.4 V
V_{out} = the difference between V_{out} measured in steps 8 and 9

11. In the FM modulator, set the dc bias voltage (V_C) to 0 V dc and adjust the amplitude of the signal generator for 0.4 V p–p and set its frequency to 1 kHz.

12. Measure the peak-to-peak demodulator output voltage (V_{out}).

13. Calculate the dynamic circuit gain using the following formula.

$$A_d = \frac{V_{out}}{V_{in}}$$

where A_d = dynamic circuit gain
V_{in} = modulator input voltage (0.4 V p–p)
V_{out} = peak-to-peak demodulated voltage (V_{out} measured in step 12)

14. Compare the dynamic circuit gain calculated in step 13 to the static circuit gain calculated in step 10.

SECTION C   FM Modulation and Demodulation

In this section the processes of FM modulation and demodulation are observed. The frequency deviation (ΔF) of an FM signal is directly proportional to the amplitude of the modulating signal and the deviation sensitivity (K_1) of the modulator. Conversely, in an FM demodulator, the output signal is directly proportional to the input frequency deviation and the frequency-to-voltage conversion gain of the demodulator. Consequently, the circuit gain is the product of the modulating signal voltage, the deviation sensitivity of the modulator, and the frequency-to-voltage conversion gain of the receiver. The FM modulator and demodulator circuits used in this section are the same as the circuits used in Section B and shown in Figure 22-3.

Procedure

1. Construct the FM modulator and demodulator circuits shown in Figures 22-3a and 22-3b.

2. In the modulator, adjust the dc bias supply voltage (V_C) and signal generator output voltage to 0 V dc and measure the modulator carrier rest frequency (F_c).

3. In the demodulator, disconnect R_1 and adjust R_x for a VCO free-running frequency equal to the modulator carrier rest frequency ($F_o = F_c$).

4. Reconnect R_1 and connect a jumper wire between the modulator output and the demodulator input.

5. Adjust the dc bias supply (V_C) to +0.2 V dc and measure the deviated modulator output frequency (F_c').

6. Calculate the frequency deviation using the following formula.

$$\Delta F = |F_c - F_c'|$$

where ΔF = frequency deviation
F_c = carrier rest frequency
F_c' = deviated output frequency

7. Calculate the modulator deviation sensitivity using the following formula.

$$K_1 = \frac{\Delta F}{\Delta V}$$

where K_1 = deviation sensitivity
ΔF = frequency deviation
ΔV = change in V_C (0.2 V)

8. Calculate the demodulator output voltage for a modulating signal voltage V_a = 0.3 V p using the following formula.

$$V_{out} = (V_a)(K_1)(K_t)$$

where V_{out} = peak demodulated voltage
V_a = modulating signal voltage
K_1 = modulator deviation sensitivity
K_t = demodulator frequency-to-voltage conversion gain (Section A, step 12)

9. In the modulator, set the dc bias voltage (V_C) to 0 V, and adjust the amplitude of the signal generator output voltage to 0.3 V p and set the frequency to 1 kHz.

10. Sketch the FM waveform at the output of the modulator.

11. Measure the demodulator output voltage (V_{out}) and compare its value to the voltage calculated in step 8.

12. Calculate the voltage gain of the PLL's internal op amp using the following formula.

$$A_v = \frac{R_C + R_F}{R_C}$$

where A_v = op amp voltage gain
R_C = inverting input resistor
R_F = feedback resistor

13. Measure the phase comparator output voltage (V_d) and determine the voltage gain of the PLL's internal op amp using the following formula.

$$A_v = \frac{V_{out}}{V_{in}}$$

where A_v = voltage gain
V_{out} = op amp output voltage
V_{in} = phase-comparator output voltage (V_d)

14. Compare the voltage gain calculated in step 12 to the voltage gain measured in step 13.

15. Describe any differences in the waveform observed at the output of the phase comparator (V_d) and the waveform observed at the output of the op amp (V_{out}).

16. Vary the frequency of the signal generator and describe what effect varying it has on the demodulated signal (V_{out}).

17. Vary the amplitude of the signal generator and describe what effect varying it has on the demodulated signal (V_{out}).

SECTION D   Summary

Write a brief summary of the concepts presented in this experiment on FM demodulators. Include the following items:

1. The concepts of frequency modulation, frequency deviation, and deviation sensitivity.
2. The concept of frequency-to-voltage conversion.
3. The concept of circuit gain.
4. The process of FM demodulation.
5. The operation of a PLL frequency demodulator.

EXPERIMENT 22   Answer Sheet

NAME: _____ CLASS: _____ DATE: _____

SECTION A

6. V_{out} = _____   7. V_{out} = _____

8. K_+ = _____   9. V_{out} = _____

10. K_- = _____   11. K_t = _____

SECTION B

2. F_c (calculated) = _____   4. F_c (measured) = _____

5. F_o (calculated) = _____   8. V_{out} = _____

9. V_{out} = _____   10. A_s = _____

12. V_{out} = _____   13. A_d = _____

14. _____

_____

_____

SECTION C

2. F_c (measured) = _____   5. $F_c{'}$ (measured) = _____

6. ΔF = _____   7. K_1 = _____

8. V_{out} = _____

10.

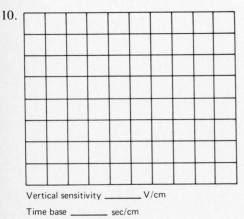

Vertical sensitivity _____ V/cm

Time base _____ sec/cm

11. V_{out} = _____

12. A_v (calculated) = _____   13. A_v (measured) = _____

14. _____
_____
_____

15. _____
_____
_____

16. _____
_____
_____

17. _____
_____
_____

TRANSMISSION LINE CHARACTERISTIC IMPEDANCE

REFERENCE TEXT: *Fundamentals of Electronic Communications Systems.*

1. Chapter 9, Uniformly Distributed Lines.
2. Chapter 9, Transmission Characteristics, Characteristic Impedance.

OBJECTIVES

1. To observe the effects of lumped transmission line parameters.
2. To determine the characteristic impedance of a transmission line.
3. To observe the characteristics of matched and mismatched transmission lines.

INTRODUCTION

The characteristics of a transmission line are determined by its electrical properties (such as wire conductivity and insulator dielectric constant) and its physical properties (such as wire diameter and conductor spacing). These properties, in turn, determine the primary electrical constants: series dc resistance (R), series inductance (L), shunt capacitance (C), and shunt conductance (G). The primary constants are uniformly distributed throughout the length of the transmission line and are, therefore, commonly called distributed parameters. To simplify analysis, distributed parameters are commonly lumped together per a given unit length to form an artificial electrical model of the line. For example, series resistance is generally given in ohms per mile or kilometer. The transmission characteristics of a transmission line are called secondary constants. The secondary constants are determined from the four primary constants. The secondary constants are characteristic impedance and propagation constant. The characteristics and behavior of a transmission line are difficult to analyze without using sophisticated and very expensive test equipment. Therefore, in this experiment, the four primary constants for a unit length of transmission line are replaced with a single series resistance (R) and a single shunt resistance (R_s). The series resistance represents the series inductance and dc resistance of the transmission line, and the shunt resistance represents the shunt capacitance and conductance. Figure 23-1a shows a source and a load interconnected by a single unit length of transmission line. The figure shows the electrical equivalent circuit for a metallic two-wire trans-

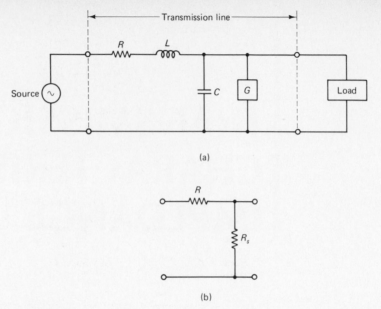

FIGURE 23-1 (a) Transmission line electrical equivalent circuit. (b) Experimental equivalent circuit.

mission line and the relative placement of the various lumped parameters. Figure 23-1b shows a single unit length of the transmission line equivalent circuit used in this experiment.

MATERIALS REQUIRED

Equipment:

1 — protoboard
1 — digital multimeter (dc resistance and ac current)
1 — audio signal generator
1 — standard oscilloscope (10 MHz)
1 — resistor substitution box

Parts List:

10 — 100-ohm resistors
10 — 1 k-ohm resistors

SECTION A   Transmission Line Characteristic Impedance

In this section the characteristic impedance of a transmission line is examined. The characteristic impedance of a transmission line (Z_o) is a complex ac quantity which is expressed in ohms, is totally independent of both length and frequency, and cannot be directly measured. Characteristic impedance (which is sometimes called surge impedance) is defined as the impedance seen looking into an infinitely long line or the impedance seen looking into a finite length of line that is terminated in a purely resistive load with a resistance equal to the characteristic impedance of the line. In this section the four primary constants of a transmission line (series resistance, series inductance, shunt capacitance,

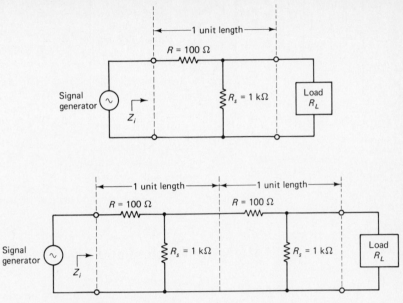

FIGURE 23-2 (a) Transmission line equivalent circuit (1 unit length).
(b) Transmission line equivalent circuit (2 unit lengths).

and shunt conductance) are simulated with a single series resistance (R) and a single shunt resistance (R_s) per unit length. The electrical equivalent circuit for a single unit length of the purely resistive transmission line used in this section is shown in Figure 23-2a. The Figure shows a source and a single section (unit length) of transmission line terminated in an open circuit.

Procedure

1. Construct the transmission line equivalent circuit shown in Figure 23-2a (leave the signal generator and load disconnected).
2. Calculate the input impedance (Z_i) of the transmission line equivalent circuit shown in Figure 23-2a.
3. Calculate the input impedance of ten sections of transmission line.
4. With an ohmmeter, measure the input impedance of the single section of transmission line shown in Figure 23-2a.
5. Add a second section of transmission line (i.e., an additional R and R_s) as shown in Figure 23-2b and measure the input impedance of the equivalent transmission line.
6. Simulate an infinitely long line by adding eight additional sections, measuring the input impedance after the addition of each section.
7. From the measurements made in step 6 determine the characteristic impedance of the simulated transmission line.
8. Terminate the simulated transmission line with a resistance substitution box and set the resistance of the box (Z_L) to the characteristic impedance determined in step 7.
9. Measure the input impedance of the equivalent transmission line and compare its value to the impedance calculated in step 3 and the last impedance measurement made in step 6.
10. Remove nine sections of transmission line one section at a time, measuring the input impedance after each section is removed (leave the substitution box connected).
11. Compare the last impedance measured in step 10 to the last impedance measured in step 6 and the impedance calculated in step 3.

12. Connect the signal generator to the remaining section of transmission line (again, leave the resistance box connected).

13. Set the amplitude of the signal generator output voltage to 10 V p–p and set its frequency to 1 kHz.

14. Calculate the transmission line current using the following formula.

$$I_o = \frac{E_o}{Z_o}$$

where I_o = rms line current
Z_o = characteristic impedance
E_o = rms source voltage

15. Measure the total ac current and compare it to the value calculated in step 14.

SECTION B   Matched and Mismatched Transmission Lines

In this section the characteristics of matched and mismatched transmission lines are examined. For maximum transfer of power, a transmission line must be matched to the load (i.e., $Z_o = Z_L$). When Z_L is greater than or less than Z_o, standing waves exist on the line and all the source power is not absorbed by the load. When Z_L is not equal to Z_o, the line is said to be mismatched and standing waves exist on the line. The transmission line equivalent circuit used in this section is shown in Figure 23-3.

Procedure

1. Construct the transmission line equivalent circuit shown in Figure 23-3 (set the resistance of the substitution box (Z_L) to the characteristic impedance (Z_o) determined in Section A).

2. Set the amplitude of the signal generator output voltage to 10 V p–p and set the frequency to 1 kHz.

3. Measure the voltage across R_L and calculate the rms power absorbed by the load using the following formula.

$$P = \frac{E^2}{R_L}$$

where P = rms power
E = rms voltage across R_L
$R_L = Z_o$

4. Change the resistance of the substitution box to $Z_L/4$, $Z_L/2$, $2\,Z_L$, and $4\,Z_L$ and measure the voltage across Z_L for each resistance value.

5. Calculate the power absorbed by the load for each of the resistances given and voltages measured in step 4.

6. Construct a graph showing the load power-versus-load resistance for the resistances given and powers calculated in steps 3 and 4.

7. Describe the relationship between the load resistance and the load power.

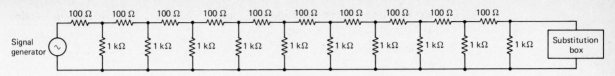

FIGURE 22-3   Transmission line equivalent circuit

SECTION C   Summary

Write a brief description of the concepts presented in this experiment on transmission lines. Include the following items:

1. The primary constants of a transmission line.
2. The equivalent circuit for a transmission line.
3. The concepts of lumped and distributed parameters.
4. The characteristic impedance of a transmission line.
5. The characteristics of a matched and mismatched line.
6. The concept of maximum power transfer to the load.

EXPERIMENT 23   Answer Sheet

NAME: _____ CLASS: _____ DATE: _____

SECTION A

2. Z_i (calculated) = _____

3. Z_i (calculated) = _____

4. Z_i (measured) = _____

5. Z_i (measured) = _____

Sections	Z_i
1	
2	
3	
4	
5	

Sections	Z_i
6	
7	
8	
9	
10	

7. Z_o (calculated) = _____

9. Z_o (measured) = _____

10.

Sections	Z_i
10	
9	
8	
7	
6	

Sections	Z_i
5	
4	
3	
2	
1	

11. _____

_____

_____

14. I_o (calculated) = _____

15. I_o (measured) = _____

SECTION B

3. P (calculated) = _____

4.

R_L	V
$Z_L/4$	
$Z_L/2$	
$2 Z_L$	
$4 Z_L$	

5.

R_L	P
$Z_L/4$	
$Z_L/2$	
$2 Z_L$	
$4 Z_L$	

6.

R_L

Power

7. _____

_____

_____

Experiment 24

FREQUENCY SHIFT KEYING MODULATOR

REFERENCE TEXT: Electronic Communications Systems: Fundamentals through Advanced

1. Chapter 2, Large-Scale-Integration Oscillators.
2. Chapter 5, Voltage-Controlled Oscillator.
3. Chapter 7, Linear Integrated-Circuit Direct FM Generator.
4. Chapter 13, Frequency Shift Keying.

OBJECTIVES

1. To observe the operation of a linear integrated-circuit function generator.
2. To observe the operation of a frequency shift keying modulator.
3. To observe the operation of a linear integrated-circuit frequency shift keying modulator.

INTRODUCTION

Frequency shift keying (FSK) is a relatively simple, low-performance form of digital modulation. FSK is a constant-envelope form of angle modulation similar to conventional frequency modulation except that the modulating signal varies between two discrete voltage levels (i.e., 1s and 0s) rather than with a continuously changing waveform. Binary FSK is the most common form of FSK. With binary FSK, the center or carrier frequency is shifted (deviated) by the binary input data. Consequently, the output of an FSK modulator is a step function in the frequency domain. As the binary input signal changes from a logic 0 to a logic 1 and vice versa, the FSK output shifts between two frequencies: a mark or logic 1 frequency and a space or logic 0 frequency. In this experiment the XR-2206 monolithic function generator is used for the FSK modulator. The block diagram for the XR-2206 function generator is shown in Figure 24-1.

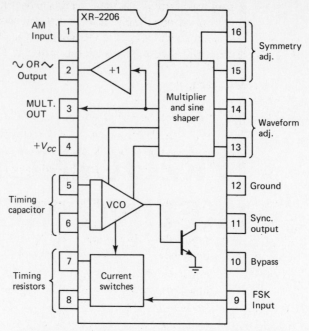

FIGURE 24-1   XR-2206 function generator block diagram.

MATERIALS REQUIRED

Equipment:

1 — protoboard

1 — dual dc power supply (+ 12 V dc and − 1 to + 3 V dc)

1 — function generator (100 kHz)

1 — standard oscilloscope (10 MHz)

1 — assortment of test leads and hook up wire

Parts list:

1 — XR-2206 monolithic function generator

1 — 3.9k-ohm resistor

3 — 4.7k-ohm resistors

1 — 6.8k-ohm resistor

1 — 10k-ohm resistor

1 — 47k-ohm resistor

1 — 1k-ohm variable resistor

1 — 10k-ohm variable resistor

2 — 0.001-UF capacitors

2 — 1-UF capacitors

1 — 10-UF capacitor

SECTION A   Frequency Shift Keying — the Sweep Mode

In this section the XR-2206 monolithic function generator operating in the sweep mode is used to generate FSK. In the sweep mode the XR-2206 function generator output frequency is proportional to the input voltage. However, with FSK, the input voltage is a binary waveform that simply changes between two discrete voltage levels. Therefore, the output frequency simply shifts or deviates between two frequencies (a mark and a space frequency) with changes in the input voltage. The schematic diagram for the FSK modulator circuit used in this section is shown in Figure 24-2.

Procedure

1. Construct the FSK modulator circuit shown in Figure 24-2 (set R_2 to midrange).
2. With the dc bias voltage (V_C) set to 0 V dc, adjust R_3 until a sine wave with minimum distortion is observed at V_{out}.
3. Adjust R_2 until an unmodulated carrier frequency $F_C = 100$ kHz is observed at V_{out}.
4. Adjust the dc bias voltage (V_C) to $+ 1$ V dc (logic 1), and record the output mark frequency.
5. Adjust the dc bias voltage (V_C) to $- 1$ V dc (logic 0), and record the output space frequency.
6. Replace the dc bias supply with a function generator, and set the output of the function generator to a 25 kHz 2 V p-p square wave (i.e., ± 1 V p).
7. Observe the FSK waveform at V_{out} (it may be necessary to adjust the function generator output frequency slightly to achieve a stable FSK waveform).
8. Sketch the digital waveform observed at the function generator output and the FSK waveform observed at the output of the FSK modulator (V_{out}).
9. Describe the two waveforms sketched in step 8.
10. Measure the FSK mark and space frequencies, and calculate the modulation index using the following formula.

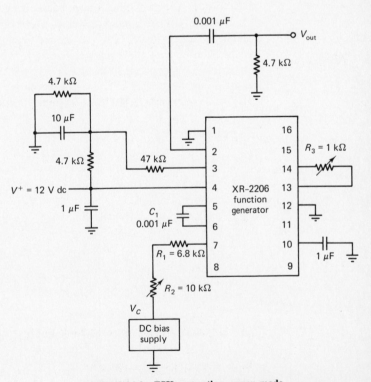

FIGURE 24-2 FSK generation, sweep mode.

$$m = \frac{|F_m - F_s|}{F_b}$$

where m = modulation index
F_m = mark frequency
F_s = space frequency
F_b = binary input data
rate (25 kbps)

11. Using the Bessel table, determine the number of significant side frequencies, and sketch the output spectrum for the FSK waveform observed in step 7.

12. What is the minimum bandwidth required to propagate the FSK signal?

13. What is the baud rate?

14. Vary the frequency of the square wave input signal, and describe what effect varying it has on the FSK output waveform.

15. Vary the amplitude of the square wave input signal, and describe what effect varying it has on the FSK output waveform.

SECTION B   Frequency Shift Keying — the Timing Resistor Mode

In this section the XR-2206 function generator operating in the timing resistor mode is used to generate FSK. The schematic diagram for the FSK modulator circuit used in this section is shown in Figure 24-3. In the timing resistor mode, the XR-2206 is operated with two separate timing resistors (R_1 and R_2) connected to timing pins 7 and 8, respectively. Depending on the level of the input signal voltage on pin 9, either R_1 or R_2 is activated. If pin 9 is open circuited or connected to a voltage greater than the input switching voltage, R_1 is activated. If pin 9 is connected to a voltage less than the input switching voltage, R_2 is activated. The switching voltage varies between 1 and 2 V depending on the specific chip, with 1.4 V being a typical value. Therefore, if a binary digital signal varying above and below approximately 1.4 V is applied to pin 9, the output signal is keyed between two frequencies (a mark and a space frequency). In essence, the VCO free-running frequency is switched (as opposed to deviated) between the mark and space frequencies at a rate proportional to the rate of change of the binary input signal. This method of generating FSK is sometimes referred to as frequency switch keying.

Procedure

1. Construct the FSK modulator circuit shown in Figure 24-3.

2. With the dc bias voltage (V_C) set to 0 V dc, adjust R_3 until a sine wave with the minimum distortion is observed at V_{out}.

3. Calculate the mark and space frequencies using the following formulas.

$$F_m = \frac{1}{R_1 C_1} \qquad\qquad F_s = \frac{1}{R_2 C_1}$$

where F_m = mark frequency
F_s = space frequency

4. Determine the actual mark frequency by setting the dc bias voltage (V_C) to $+3$ V dc and measuring the frequency of the waveform at V_{out}.

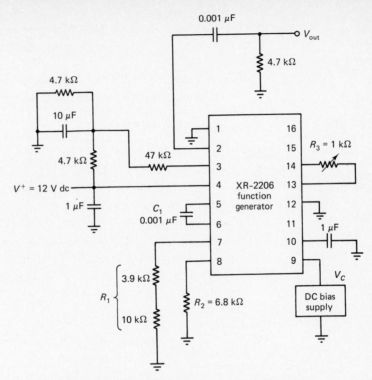

FIGURE 24-3 FSK generation, timing resistor mode.

5. Determine the actual space frequency by setting the dc bias voltage (V_C) to 0 V dc and measuring the frequency of the waveform at V_{out}.

6. Determine the XR-2206 switching voltage by slowly increasing V_C until the output switches from the space to the mark frequency.

7. Replace the dc bias supply with a function generator, and set the amplitude of the function generator output voltage to 3 V p-p (a maximum positive voltage of $+3$ V and a minimum positive voltage of 0 V). Set the signal generator output frequency to a 20 kHz square wave.

8. Observe the FSK waveform at V_{out} (it may be necessary to adjust the function generator output frequency slightly to achieve a stable FSK waveform).

9. Sketch the binary waveform observed at the function generator output and the FSK waveform observed at the output of the FSK modulator (V_{out}).

10. Describe the two waveforms sketched in step 9.

11. What is the baud rate?

12. Vary the frequency of the square wave input signal, and describe what effect varying it has on the FSK output waveform.

13. Vary the amplitude of the square wave input signal, and describe what effect varying it has on the FSK output waveform.

SECTION C Summary

Write a brief summary of the concepts presented in this experiment on FSK modulators. Include the following items:

1. The basic concept of FSK modulation.

2. The sweep frequency mode of generating FSK.

3. The timing resistor mode of generating FSK.

4. The relationship between the amplitude of the binary input voltage and the FSK waveform in both the sweep frequency and timing resistor modes of FSK generation.

5. The relationship between the frequency of the binary input voltage and the FSK waveform in both the sweep frequency and timing resistor modes of FSK generation.

6. The relationship between the bit and baud rates with FSK.

NAME: _____ CLASS: _____ DATE: _____

SECTION A

4. Mark frequency = _____ 5. Space frequency = _____

8.

(grid)	(grid)

Vertical sensitivity _____ V/cm   Vertical sensitivity _____ V/cm

Time base _____ sec/cm   Time base _____ sec/cm

9. _____

_____

_____

10. Mark frequency = _____, Space frequency = _____, m = _____

11.

(graph: Voltage vs Frequency)

12. Minimum bandwidth = _____

13. Baud rate = _____

14. _____

_____

_____

15. _____

_____

_____

SECTION B

3. F_m = _____, F_s = _____   4. F_m = _____
5. F_s = _____   6. V_C = _____
9.

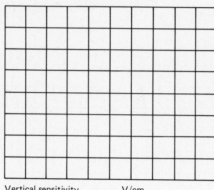

Vertical sensitivity _____ V/cm   Vertical sensitivity _____ V/cm

Time base _____ sec/cm   Time base _____ sec/cm

10. _____
_____
_____

11. Baud rate = _____

12. _____
_____
_____

13. _____
_____
_____

PHASE-LOCKED LOOP FSK DEMODULATOR

REFERENCE TEXT: Electronic Communications Systems: Fundamentals through Advanced

1. Chapter 5, Phase-Locked Loop.
2. Chapter 8, PLL FM Demodulators.
3. Chapter 13, FSK Receivers.

OBJECTIVES

1. To observe the operation of an FSK Demodulator.
2. To observe the operation of a phase-locked loop FSK demodulator.
3. To observe the operation of a linear integrated-circuit FSK demodulator.

INTRODUCTION

The most common circuit used for demodulating FSK signals is the phase-locked loop (PLL). A PLL-FSK demodulator works much like a PLL FM demodulator. As the external input frequency to the PLL shifts between a mark and a space frequency and vice versa, the dc error voltage at the output of the phase comparator changes proportionally. Because there are only two input frequencies (a mark and a space), there are only two output voltages. One output voltage represents a logic 0, and the other a logic 1. Therefore, the output is a two-level (binary) representation of the FSK input. In this experiment the XR-2212 monolithic phase-locked loop is used for the FSK demodulator, and the XR-2206 monolithic function generator is used for the FSK modulator. The functional block diagrams for the XR-2212 and XR-2206 are shown in Figure 25-1.

MATERIALS REQUIRED

Equipment:

1 — protoboard
1 — dual dc power supply (+12 V dc and −1 V dc to +1 V dc)
1 — function generator (100 kHz)

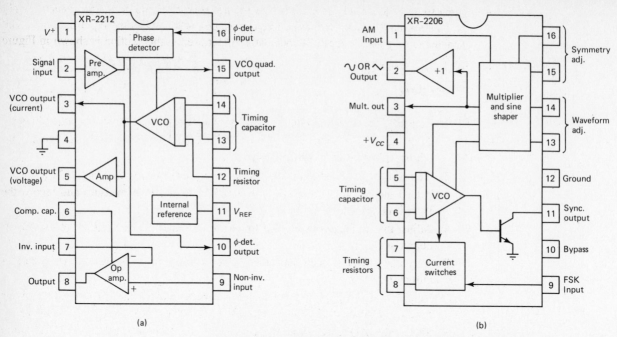

FIGURE 25-1 Functional block diagrams. (a) XR-2212 monolithic PLL. (b) XR-2206 monolithic function generator.

1 — standard oscilloscope (10 MHz)

1 — assortment of test leads and hookup wire

Parts list:

1 — XR-2206 monolithic function generator

1 — XR-2212 monolithic phase-locked loop

5 — 4.7k-ohm resistors

1 — 6.8k-ohm resistor

3 — 10k-ohm resistors

2 — 47k-ohm resistors

2 — 100k-ohm resistors

1 — 1k-ohm variable resistor

2 — 10k-ohm variable resistors

1 — 30-pF capacitor

5 — 0.001-μF capacitors

3 — 0.1-μF capacitors

2 — 1-μF capacitors

1 — 10-μF capacitor

SECTION A The FSK Modulator

In this section the output frequency-versus-input voltage characteristics of the XR-2206 monolithic FSK modulator operating in the sweep mode are examined. In the sweep mode, the XR-2206 output frequency is proportional to its input voltage. Because the input to an FSK modulator is a binary (two-level) voltage, the output frequency simply shifts between the mark and space frequencies proportionally to the input-data rate. The schematic diagram for the FSK modulator circuit used in this section is shown in Figure 25-2.

Procedure

1. Construct the FSK modulator circuit shown in Figure 25-2 (set R_2 to midrange).
2. With the dc bias voltage (V_C) set to 0 V dc, adjust R_3 until a sine wave with minimum distortion is observed at V_{out}.
3. Adjust R_2 until an unmodulated carrier $F_c = 100$ kHz is observed at V_{out}.
4. Adjust the dc bias voltage (V_C) to $+0.5$ V dc and then measure and record the output mark frequency.
5. Adjust the dc bias voltage (V_C) to -0.5 V dc and then measure and record the output space frequency.
6. Vary the dc bias voltage, and describe what effect varying it has on the modulator output waveform.
7. Replace the dc bias supply with a function generator, and set the output of the function generator to a $100 =$ Hz, 1V Vp–p square wave (i.e. $+/-0.5$ V p).
8. Sketch the FSK waveform at V_{out} (it may be necessary to adjust the function generator output frequency slightly to achieve a stable FSK waveform).
9. Measure the mark and space frequencies, and verify that they are close to the values measured in steps 4 and 5.
10. Calculate the frequency deviation for the modulator using the following formula.

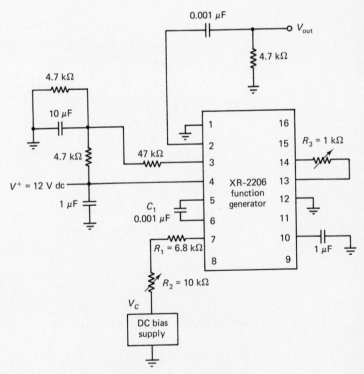

FIGURE 25-2   FSK modulator.

$$\Delta F = \frac{|F_m - F_s|}{2}$$

where ΔF = frequency deviation
F_m = mark frequency
F_s = space frequency

11. Leave this circuit assembled, because it is needed to check the operation of the FSK demodulator circuit in Section B.

SECTION B   The PLL-FSK Demodulator

In this section the operation of a PLL-FSK demodulator is examined. Once frequency lock has occurred, the output voltage from a PLL is proportional to the difference between the external input frequency and the VCO free-running frequency. Therefore, as the external FSK input shifts between a mark and a space frequency and vice versa, the PLL output voltage varies above and below its reference output voltage. The schematic diagram for the PLL-FSK demodulator circuit used in this section is shown in Figure 25-3.

Procedure

1. Construct the PLL-FSK demodulator circuit shown in Figure 25-3.
2. Disconnect one end of resistor R_1, and adjust R_x until the VCO free-running frequency is equal to 100 kHz.

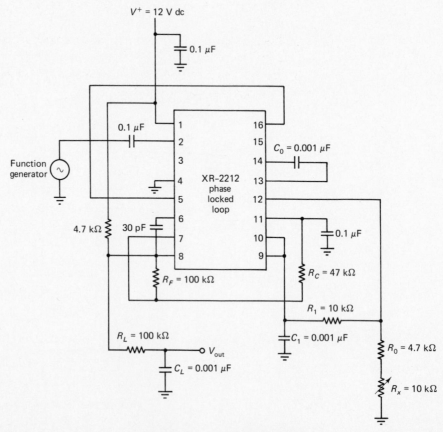

FIGURE 25-3   PLL-FSK Demodulator.

3. Reconnect R_1, and set the amplitude of the function generator output voltage to a 2 Vp–p sine wave and set its frequency to the modulator space frequency measured in step 5 of Section A.
4. Measure the dc voltage at V_{out}.
5. Change the function generator output frequency to the modulator mark frequency measured in step 4 of Section A.
6. Measure the dc voltage at V_{out}.
7. Calculate the peak-to-peak difference in the PLL output voltages measured in steps 4 and 6.
8. Connect the function generator to the FSK modulator, and set its output to a 100-Hz, 1 Vp–p square wave (i.e., $\pm$ V p).
9. Connect the output from the FSK modulator to the input to the FSK demodulator, and measure the peak-to-peak demodulator output voltage.
10. Compare the output voltage measured in step 9 to the voltage calculated in step 7.
11. Vary the function generator output frequency, and describe what effect varying it has on the demodulator output signal.
12. Vary the function generator output voltage, and describe what effect varying it has on the demodulator output signal.

SECTION C   Summary

Write a brief summary of the concepts presented in this experiment on FSK demodulators. Include the following items:

1. The relationship between the FSK demodulator input frequency and its output voltage.
2. The relationship between the FSK modulator input frequency and the FSK-demodulator output voltage.
3. The relationship between the amplitude of the FSK modulator input voltage and the FSK demodulator output voltage.

EXPERIMENT 25 ANSWER SHEET

NAME: _____ CLASS: _____ DATE: _____

SECTION A

4. Mark frequency = _____   5. Space frequency = _____

6. _____

_____

_____

8.

Vertical sensitivity _____ V/cm

Time base _____ sec/cm

9. Mark frequency = _____, Space frequency = _____

10. Frequency deviation = _____

SECTION B

4. V_{out} = _____   6. V_{out} = _____

7. V_{out} = _____   9. V_{out} = _____

10. _____

_____

_____

11. _____

_____

_____

12. _____

_____

_____

Experiment 26

FSK DEMODULATOR TONE DETECTOR

REFERENCE TEXT: Electronic Communications Systems: Fundamentals through Advanced

1. Chapter 5, Phase-Locked Loop.
2. Chapter 8, PLL FM Demodulators.
3. Chapter 13, FSK Receivers.

OBJECTIVES

1. To observe the operation of an FSK demodulator.
2. To observe the operation of a phase-locked loop FSK demodulator.
3. To observe the operation of a special-purpose linear integrated-circuit FSK demodulator.
4. To observe the operation of a carrier (tone) detect circuit.

INTRODUCTION

Recently, special-purpose monolithic integrated circuits have been designed for FSK demodulation. The XR-2211 is a monolithic phase-locked loop (PLL) system especially designed for data communications. The XR-2211 is particularly well suited for FSK demodulation because it operates over a wide supply-voltage range (4.5 to 20 V dc) and a wide frequency range (0.01 Hz to 300 kHz). The XR-2211 can accommodate analog input signals between 2 mV and 3 V, and it can interface with conventional DTL, TTL, and ECL logic families. The XR-2211 consists of a basic band limited PLL for tracking, a quadrature phase detector (with carrier detection), and an FSK voltage comparator which provides FSK demodulation. External components are used to independently set the free-running frequency of the voltage-controlled oscillator (VCO), the loop bandwidth, and the output delay. The main PLL within the XR-2211 is constructed from an input pre-amplifier, an analog multiplier used as a phase detector, and a precision VCO. The functional block diagram for the XR-2211 is shown in Figure 26-1.

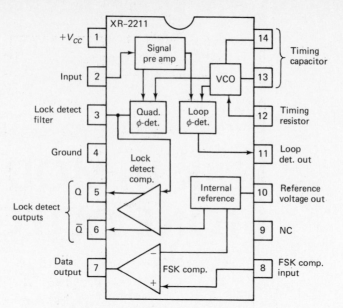

FIGURE 26-1   XR-2211 Functional Block Diagram.

MATERIALS REQUIRED

Equipment:

1 — protoboard
1 — dc power supply (12 V dc)
1 — signal generator (1 MHz)
1 — standard oscilloscope (10 MHz)
1 — assortment of test leads and hookup wire

Parts list:

1 — XR-2211 monolithic FSK demodulator/tone detector
1 — 4.7k-ohm resistor
1 — 100k-ohm resistor
1 — 470k-ohm resistor
3 — 0.1-UF capacitors

1 — 10k-ohm variable resistor
1 — assortment of resistors and capacitors whose values depend on the circuit design

SECTION A   FSK Demodulator Design Equations

In this section the design equations for the XR-2211 FSK demodulator are examined. The input preamplifier in the XR-2211 is used as a limiter such that input signals above approximately 2 mV rms are amplified to a constant high signal level. The multiplying-type phase detector acts as a digital exclusive-OR gate. Its unfiltered output produces the sum and difference frequencies of the external FSK input signal and the VCO output frequency (i.e., once frequency lock has occurred, $2 \times F$in and 0 Hz). The sum frequency is removed with filtering, and the dc component (0 Hz) drives the VCO. The VCO is actually a current-controlled oscillator with its nominal input current (I_0) set by external

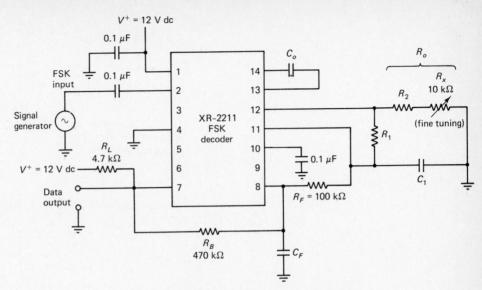

FIGURE 26-2   XR-2211 FSK demodulator.

resistor (R_0) to ground and its driving current set by external resistor (R_1) from the phase detector. The schematic diagram for the FSK demodulator circuit used in this section is shown in Figure 26-2. For the FSK demodulator designed in this section, use mark and space frequencies of 30 and 50 kHz, respectively, and a bit rate $F_b = 50$ bps.

Procedure

1. Calculate the VCO free-running frequency using the following formula.

$$F_o = \frac{|F_m + F_s|}{2}$$

    where F_o = VCO free-running frequency
               F_m = mark frequency
               F_s = space frequency

2. Calculate the external timing capacitance value using the following equation.

$$C_o = \frac{1}{F_o R_o}$$

    where C_o = external timing capacitance
               R_o = external timing resistance (an
                     arbitrary value between 10 and
                     100 kΩ; use 25 kΩ)
               F_o = VCO free-running frequency

3. Calculate the peak-to-peak frequency deviation using the following formula.

$$\Delta F = |F_m - F_s|$$

    where ΔF = frequency deviation (peak-to-peak)
               F_m = mark frequency
               F_s = space frequency

4. Calculate the resistance value for the PLL external lowpass filter using the following formula.

$$R_1 = R_o \frac{F_o}{(\Delta F)}$$

where
R_1 = lowpass filter resistance
F_o = VCO free-running frequency
R_o = external timing resistance
ΔF = frequency deviation

5. Calculate the capacitance value for the PLL external lowpass filter with a damping factor $\tau = 0.5$ using the following formula.

$$C_1 = \frac{C_o}{4}$$

where
C_1 = lowpass-filter capacitance
C_o = external timing resistance

6. Calculate the data-filter capacitance using the following formula.

$$C_F = \frac{3}{(\text{bit rate})} \ \mu F$$

SECTION B   FSK Demodulator Implementation

In this section the FSK demodulator circuit designed in Section A is constructed and examined. The FSK demodulator circuit shown in Figure 26-2 is used for this section.

Procedure

1. Construct the FSK demodulator circuit shown in Figure 26-2 (use the resistance and capacitance values calculated in Section A, and use a 10k ohm variable resistor in series with R_2 for fine tuning).
2. Measure the VCO free-running frequency at pin 3 with no external input signal and with pin 10 shorted to pin 2.
3. Adjust R_x for a VCO free-running frequency of 40 kHz.
4. Remove the short from between pins 2 and 10.
5. Connect the output of the signal generator to the PLL-FSK input, and set the amplitude of the signal generator output voltage to a 2 Vp-p sine wave with a frequency of 50 kHz.
6. Measure the data output voltage on pin 7.
7. Change the signal generator output frequency to 30 kHz, and measure the data output voltage on pin 7.
8. Measure the frequency at which the data output voltage level changes from a mark (logic 1) to a space (logic 0) level by slowly increasing the signal generator output frequency from 30 kHz to 50 kHz.

SECTION C   FSK Decoding with Carrier Detect

In this section the lock-detect section of the XR-2211 is used as a carrier-detect circuit. The schematic diagram for the circuit used in this section is shown in Figure 26-3. The open collector lock-detect output (pin 6) is shorted to the data output (pin 7). Consequently, the data output is automatically disabled or held in the low state in the absence of a received carrier. This feature of the XR-2211 is ideally suited for the receive-line-signal detect (carrier-detect) circuit when using the FSK demodulator with the RS 232C interface. The minimum value for the lock-detect capacitor (C_D) is inversely proportional to the PLL capture range ($\pm \triangle F$), which is the range of acceptable receive frequencies over which the loop can acquire lock. If a value for C_D is selected that is below the minimum capacitance, chatter is observed on the lock-detect output as the frequency of the incoming signal approaches the capture bandwidth. Excessively large values for C_D will slow the response time of the lock-detect output.

Procedure

1. Construct the FSK demodulator circuit shown in Figure 26-3 (use the resistance and capacitance values calculated in Section A).
2. Calculate the capacitance value for C_D using the following formula.

$$C_D \; (\mu F) \geq \frac{16}{\text{capture range (Hz)}}$$

$$
\begin{aligned}
\text{where} \quad CD &= \text{lock-detect-} \\
&\quad \text{filter capacitance} \\
\text{capture} &= |F_m - F_s| = \Delta F \\
\text{range}
\end{aligned}
$$

3. Set the amplitude of the signal generator output voltage to 2 V-p-p sine wave, and adjust its frequency to 60 kHz.
4. Measure the dc voltage on the carrier-detect output (pin 3).

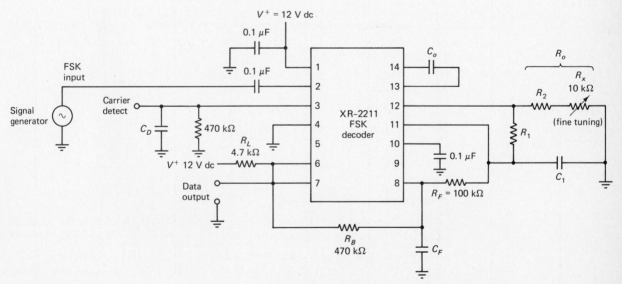

FIGURE 26-3   XR-2211 FSK demodulator with carrier detect.

5. Measure the data output voltage on pin 7.
6. Slowly decrease the signal generator output frequency until the carrier-detect voltage on pin 3 changes significantly.
7. Measure the carrier-detect voltage on pin 3 and the signal-generator output frequency (F_U).
8. Decrease the signal generator output frequency until the carrier-detect voltage returns to the value measured in step 4.
9. Measure the carrier-detect voltage on pin 3 and the signal-generator output frequency (F_L).
10. Determine the PLL capture range using the following formula.

$$\Delta F = F_U - F_L$$

where ΔF = PLL Capture range
F_U = upper capture limit
F_L = lower capture limit

SECTION D   Summary

Write a brief summary of the concepts presented in this experiment on FSK demodulators. Include the following items:

1. The basic operation of the XR-2211 monolothic FSK demodulator
2. The relationship between the FSK demodulator input frequency and its output voltage
3. The basic operation of a carrier-detect circuit
4. The concept of PLL capture range

EXPERIMENT 26 ANSWER SHEET

NAME: _____ CLASS: _____ DATE: _____

SECTION A

1. F_O = _____

2. C_O = _____

3. $\triangle F$ = _____

4. R_1 = _____

5. C_1 = _____

6. C_F = _____

SECTION B

2. VCO free-running frequency = _____

6. V_{out} = _____

7. V_{out} = _____

8. F = _____

SECTION C

2. C_D = _____

4. V_{out} = _____

5. V_{out} = _____

7. V_{out} = _____

9. V_{out} = _____

10. $\triangle F$ = _____

FSK TRANSMITTER
RECEIVER DESIGN

REFERENCE TEXT Electronic Communications Systems: Fundamentals through Advanced

1. Chapter 2, Large Scale-Integration Oscillators
2. Chapter 5, Phase-Locked Loop.
3. Chapter 5, Voltage-Controlled Oscillater.
4. Chapter 7, Linear Integrated Circuit Direct FM Generator.
5. Chapter 8, PLL FM Demodulator.
6. Chapter 13, FSK Receivers.
7. Chapter 13, Frequency-Shift Keying.
8. Chapter 14, Asynchronous Modems.

OBJECTIVES

1. To design, construct, test, and troubleshoot an FSK transmitter to a given set of specifications.
2. To design, construct, test, and troubleshoot an FSK receiver to a given set of specifications.
3. To design, construct, test, and troubleshoot an FSK transmitter/receiver system to a given set of specifications.

INTRODUCTION

In this experiment a complete operational FSK transmitter/receiver system will be designed, constructed, and tested to a given set of specifications. The FSK transmitter will be constructed using the XR-2206 monolithic function generator as the basic building block, and the FSK receiver will be constructed using either the XR-2212 monolithic phase-locked loop or the XR-2211 monolithic FSK demodulator/tone detector as the basic building block. Design equations and examples that are given in the manufacturer's specification sheets for the three integrated circuit special function chips will be used to determine the basic circuit configurations and to calculate specific component values. The simplified block diagram for the FSK transmitter/receiver system that will be designed in this experiment is shown in Figure 27-1.

211

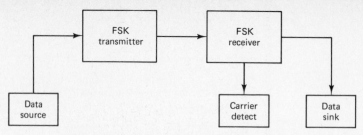

FIGURE 27-1   FSK Transmitter/receiver system.

MATERIALS REQUIRED

Equipment:

1 — protoboard

1 — dual dc power supply (12 V dc and +6 V dc to −6 V dc)

1 — signal generator (100 kHz)

1 — function generator (100 kHz)

1 — standard oscilloscope (10 MHz)

1 — assortment of test leads and hookup wire

Parts list:

1 — XR-2206 monolithic function generator

1 — XR-2212 monolithic phase-locked loop or one XR-2211 monolithic FSK
demodulator/tone detector

1 — assortment of resistors and capacitors whose value depends on the particular
design requirements

SECTION A   Design Specifications

In this section the design specifications for the FSK transmitter/receiver system that will
be designed in this experiment are given. The system must be designed to operate in the
simplex (one-way only) mode. Therefore, only one FSK transmitter and one FSK receiver
will be constructed and tested. The FSK system specifications for this experiment will
meet the manufacturer's specifications for the Bell System 103 modem operating in either
the originate or answer modes. The receiver must include a visual indication of carrier
detect (such as an LED). The binary signal source can be simulated with a function
generator. The specifications for the Bell System 103 modem are given in Table 27-1.

TABLE 27-1   Bell System 103 Modem Specifications

| | | Operating frequencies | |
		Originate mode	Answer mode
Center frequency		1170 Hz	2125 Hz
Mark frequency		1270 Hz	2225 Hz
Space frequency		1070 Hz	2025 Hz
Operating mode:	simplex		
Timing:	asynchronous		
Modulation:	frequency-shift keying		
Bit rate:	300 bps		

SECTION B   FSK Transmitter Design

In this section the design equations given in the XR-2206 specification sheet are used to determine specific component values and operating parameters for the FSK transmitter.

Procedure

1. Calculate the following system parameters: frequency deviation, modulation index, number of significant sideband pairs, and output voltage spectrum.
2. Calculate the following transmitter-component values:

$$\frac{\text{sweep mode}}{R_1, C_1} \qquad \frac{\text{timing resistor mode}}{C_1, R_1, R_2}$$

3. Construct the FSK modulator circuit designed in steps 1 and 2.
4. Test the FSK modulator circuit constructed in step 3 to determine if the modulator is operating within the design specifications outlined in Section A.
5. Troubleshoot the FSK modulator as required, and make any design changes, adaptions, or additions that are necessary for the transmitter to operate within the given specifications.

SECTION C   FSK Receiver Design

In this section the design equations given in the XR-2212 or XR-2211 specification sheets are used to determine the specific component values and operating parameters for the FSK receiver.

Procedure

1. Calculate the following receiver-component values and system parameters.

   Component values: R_0, C_0, R_1, R_C, and R_F
   System parameters: τ, Γ, loop tracking bandwidth, K_0, K_A, K_ϕ, and K_T

2. Construct the FSK demodulator circuit designed in step 1.
3. Test the FSK demodulator circuit constructed in step 2 to determine if the demodulator is operating within the design specifications outlined in Section A.
4. Troubleshoot the FSK demodulator circuit as required, and make any design changes, adaptions, or additions that are necessary for the demodulator to operate within the given specifications.

SECTION D   Circuit Implementation

In this section the FSK transmitter and receiver circuits designed, constructed, and tested in Sections B, and C are put together and tested as a system. Make any adjustments and changes necessary for the system to operate within the given specifications.

SECTION E Summary

Write a complete laboratory report for the FSK transmitter/receiver system designed in Sections B, C, and D. Include the following items:

1. Complete schematic diagrams of the transmitter/receiver.
2. An introduction including a statement of the problem.
3. All necessary calculations.
4. All pertinent data.
5. A list of problems encountered and how they were rectified.
6. A conclusion.

EXPERIMENT 27 ANSWER SHEET

NAME: _____ CLASS: _____ DATE: _____

SECTION B

1. Frequency deviation = _____
   Modulation index = _____
   Sideband pairs = _____

2. Sweep mode   Timing-resistor mode
   R_1 = _____ C_1 = _____
   C_1 = _____ R_1 = _____
                                      R_2 = _____

SECTION C

1. R_0 = _____ R_C = _____
   C_0 = _____ R_F = _____
   R_1 = _____
   τ = _____
   Γ = _____
   Loop-tracking bandwidth = _____
   K_0 = _____
   K_A = _____a
   K_ϕ = _____
   K_T = _____

BINARY PHASE SHIFT KEYING MODULATOR

REFERENCE TEXT: Electronic Communications Systems: Fundamentals through Advanced

1. Chapter 6, Balanced Ring Modulator.
2. Chapter 13, Binary Phase Shift Keying.
3. Chapter 14, Asynchronous Modems.

OBJECTIVES

1. To observe the output phase-versus-input voltage characteristics of a balanced differential amplifier.
2. To observe the operation of a phase shift keying modulator.
3. To observe the operation of a binary phase shift keying modulator.
4. To observe the operation of a linear integrated-circuit binary phase shift keying modulator.

INTRODUCTION

Phase shift keying (PSK) is a form of angle-modulated, constant-envelope digital modulation. PSK is similar to conventional phase modulation except that with PSK the input signal is a binary digital signal and a limited number of output phases are possible. Binary phase shift keying (BPSK) is the simplest form of PSK. With BPSK, only two output phases are possible. One output phase represents a logic 1, and the other a logic 0. As the input signal voltage changes from a logic 0 to a logic 1 and vice versa, the phase of the output carrier shifts between two angles that are 180° out of phase. Another name for BPSK is phase reversal keying (PRK). In this experiment the XR-2206 monolithic function generator is used for the BPSK modulator. The functional-block diagram for the XR-2206 function generator is shown in Figure 28-1.

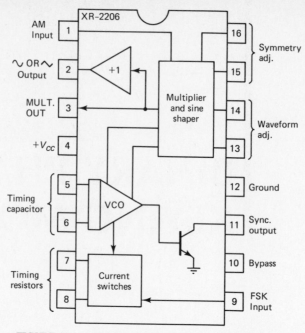

FIGURE 28-1   XR-2206 Function generator block diagram.

MATERIALS REQUIRED

Equipment:

1 — protoboard
1 — dual dc power supply (+12 V dc and 0 to +10 V dc)
1 — function generator (25 kHz)
1 — standard oscilloscope (10 MHz)
1 — assortment of test leads and hookup wire

Parts list

1 — XR-2206 monolithic function generator
4 — 4.7k-ohm resistors
1 — 6.8k-ohm resistor
1 — 10k-ohm resistor
1 — 47k-ohm resistor
1 — 1k-ohm variable resistor
1 — 10k-ohm variable resistor
2 — 0.001-μF capacitors
2 — 1-μF capacitors
1 — 10-μF capacitor

SECTION A   Output Amplitude and Output Phase versus Input Voltage

In this section both the output-amplitude and the output phase-versus-input voltage characteristics of the XR-2206 function generator are examined. The schematic diagram for the function generator circuit used in this section is shown in Figure 28-2. Pin 1 of the

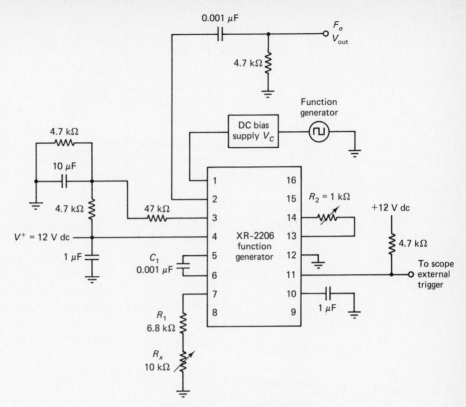

FIGURE 28-2   BPSK modulator.

XR-2206 is an input to the function generator output differential amplifier. The other input is internally biased at $V^+/2$. Therefore, the function generator output voltage can be varied by applying a dc voltage (V_C) to pin 1. As V_C increases from 0 V toward $V^+/2$, the output amplitude decreases. When $V_C = V^+/2$, the output voltage goes to approximately 0 V (V_{out} nulls); and as V_C is increased beyond $V^+/2$, the output amplitude increases except with the opposite phase. This property is suitable for both BPSK and suppressed-carrier AM.

Procedure

1. Construct the function generator circuit shown in Figure 28-2 (set V_C to 0 V dc).
2. Vary R_2 until a sine wave with minimum distortion is observed at V_{out}.
3. Adjust R_x until the output frequency $F_o = 100$ kHz.
4. Slowly increase the dc bias voltage until $V_C = V^+/2$.
5. Describe what effect increasing V_C has on the output waveform.
6. Increase the dc bias voltage until $V_C = +10$ V dc.
7. Describe what effect increasing V_C beyond $V^+/2$ has on the output waveform.
8. Set the dc supply voltage (V_C) to 0 V dc, and connect the oscilloscope external-trigger input to pin 11 of the XR-2206. (Pin 11 is a square wave output signal with a frequency equal to F_o. Synchronizing the oscilloscope to this signal establishes a phase reference for V_{out}.)
9. Calculate the dc input voltage where the output signal undergoes a 180° phase reversal using the following formula.

$$V_x = \frac{V^+}{2}$$

where V_x = input voltage where an output
                  phase reversal occurs

          V^+ = function generator supply
                  voltage

10. Slowly increase the dc supply voltage (V_C) from 0 V to $+10$ V dc while observing V_{out}.

11. What was the actual input voltage where the output signal reversed phase?

SECTION B   Binary Phase Shift Keying Modulator

In this section the operation of a BPSK modulator is examined. The same circuit that is shown in Figure 28-2 is used for the modulator except that a function generator is placed in series with the dc bias supply. The function generator simulates a binary digital input signal. The XR-2206 acts like a phase reversing switch that is controlled by the external input voltage applied to pin 1. If the input voltage has an average voltage equal to $V^+/2$ and varies above and below this value, the phase of the output signal will reverse proportionally.

Procedure

1. Construct the BPSK modulator circuit shown in Figure 28-2.

2. Set the dc bias supply voltage (V_C) to $V^+/2$.

3. Fine tune V_C until V_{out} goes to minimum (V_C should be approximately equal to $V^+/2$ and V_{out} should be very nearly 0 V).

4. Set the amplitude of the function generator output voltage to a 2 Vp-p square wave with a frequency equal to 25 kHz (i.e., the combined external input voltage should vary from $+5$ V to $+7$ V).

5. Observe the waveform at V_{out} (you may have to fine tune the function generator output frequency to observe a stable BPSK waveform).

6. Adjust V_C for a BPSK output waveform with uniform amplitude.

7. Sketch the waveform observed in step 6.

8. Describe the waveform sketched in step 7.

9. Increase the function generator output voltage to 4 Vp-p.

10. Repeat steps 5 through 8.

11. Vary the function generator output frequency, and describe what effect varying it has on V_{out}.

12. Vary the amplitude of the function generator output voltage, and describe what effect varying it has on V_{out}.

SECTION C   Summary

Write a brief summary of the concepts presented in this experiment on BPSK modulators. Include the following items:

1. The relationship between a BPSK output phase and its input-signal voltage.

2. The relationship between a BPSK output phase and its input-signal frequency.

3. The basic operation of the XR-2206 BPSK modulator.

EXPERIMENT 28 ANSWER SHEET

NAME: _____ CLASS: _____ DATE: _____

SECTION A

5. _____
   _____
   _____

7. _____
   _____
   _____

9. $V_x =$ _____   11. $V =$ _____

SECTION B

7.

Vertical sensitivity _____ V/cm

Time base _____ sec/cm

8. _____
   _____
   _____

10.

Vertical sensitivity _____ V/cm

Time base _____ sec/cm

_____

_____

11. _____

_____

_____

12. _____

_____

_____

_____

Experiment 29

BINARY PHASE SHIFT KEYING DEMODULATOR

REFERENCE TEXT: Electronic Communications Systems: Fundamentals through Advanced

1. Chapter 5, Phase-Locked Loop.
2. Chapter 8, PLL-FSK Demodulator.
3. Chapter 13, BPSK Receiver.

OBJECTIVES

1. To observe the operation of a phase shift keying demodulator.
2. To observe the operation of a binary phase shift keying demodulator.
3. To observe the operation of a linear integrated-circuit binary phase shift keying demodulator.

INTRODUCTION

The most common types of BPSK demodulators use a balanced modulator to mix the incoming BPSK signal with a coherent recovered carrier. The output from the balanced modulator is the original binary input data. In this experiment an XR-2206 function generator is used for the BPSK transmitter, and an XR-2212 phase-locked loop is used for the demodulator. The phase detector in the XR-2212 demodulator is used as a balanced product detector. The functional block diagrams for the XR-2212 monolithic phase-locked loop and XR-2206 monolithic function generator are shown in Figure 29-1.

MATERIALS REQUIRED

Equipment:

1 — protoboard
1 — dual dc power supply (+ 12 V dc and 0 to + 10 V dc)
1 — function generator (1 kHz)

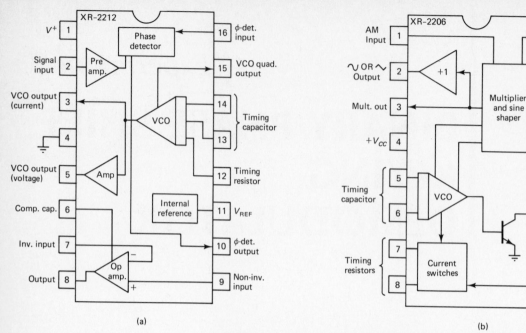

FIGURE 29-1   Functional block diagrams. (a) XR-2212 monolithic PLL. (b) XR-2206 monolithic function generator.

1 — standard oscilloscope (10 MHz)

1 — assortment of test leads and hookup wire

Parts list:

1 — XR-2206 monolithic function generator

1 — XR-2212 monolithic phase-locked loop

5 — 4.7k-ohm resistors

1 — 10k-ohm resistor

1 — 18k-ohm resistor

2 — 47k-ohm resistors

2 — 100k-ohm resistors

1 — 1k-ohm variable resistor

1 — 10k-ohm variable resistor

1 — 30-pF capacitor

5 — 0.001-μF capacitors

3 — 0.1-μF capacitors

2 — 1-μF capacitors

1 — 10-μF capacitor

SECTION A   The BPSK Modulator

In this section a BPSK modulator is constructed with the XR-2206 monolithic function generator. The schematic diagram for the BPSK modulator used in this section is shown in Figure 29-2. The carrier frequency is set with external capacitor C_1 and external resistors R_1 and R_x. The dc bias supply (V_C) is used to null the carrier, while the function generator simply supplies an alternating 1/0 binary input-data sequence. The XR-2206 acts like a phase reversing switch where the binary input data controls the transmitter output phase.

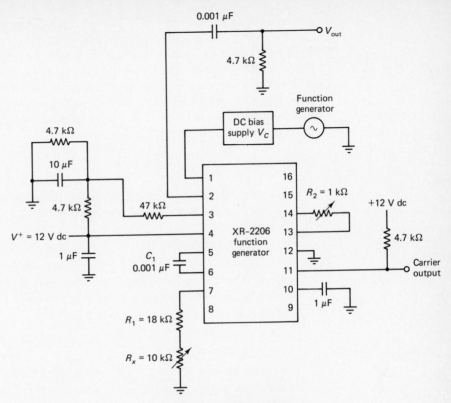

FIGURE 29-2   BPSK modulator.

Procedure

1. Construct the BPSK modulator circuit shown in Figure 29-2.
2. Set the output of the function generator to 0 V.
3. Adjust R_2 for a sine wave output with minimum distortion at V_{out}.
4. Set the dc bias supply (V_C) to 0 V dc, and adjust R_x for an output frequency $F_o = 25$ kHz.
5. Set the dc bias supply (V_C) to $V^+/2$.
6. Fine tune V_C until the output carrier is nulled (i.e., $V_{out} = 0$ V).
7. Set the amplitude of the signal generator output voltage to a 4 Vp-p square wave with a frequency equal to 100 Hz.
8. Sketch the waveform observed at V_{out} (you may have to fine tune the function generator output frequency to obtain a stable BPSK waveform).
9. Describe the waveform observed in step 8.
10. Do not disassemble this circuit, as it is needed in Section B.

SECTION B   The BPSK Demodulator

In this section the operation of a BPSK demodulator is examined. The external input to a BPSK demodulator is a sine wave with a phase angle equal to either 0° or 180° and is mixed in a product detector with a coherent recovered carrier. The output from the product detector is a sine wave with a frequency equal to the recovered carrier and a dc component ($+V$ = logic 1 and $-V$ = logic 0). The carrier component is filtered off, leaving only a dc voltage which is the recovered binary data. The BPSK demodulator circuit used in this section is shown in Figure 29-3. The PLL internal VCO circuit is not used. The PLL phase detector, which is a product modulator/demodulator, is used to demodulate the BPSK input waveform. One input to the phase detector is the BPSK input, which is connected directly to the output from the BPSK transmitter constructed in Section A. Because of the difficulty in constructing a stable carrier recovery circuit, the recovered carrier input to the phase detector is supplied directly from the square wave output from the XR-2206 function generator shown in Figure 29-2. The PLL internal op amp is used to filter and amplify the demodulated binary data.

Procedure

1. Construct the BPSK demodulator circuit shown in Figure 29-3, and connect the output from the BPSK modulator shown in Figure 29-2 to the demodulator BPSK input.

2. With the dc bias supply (V_C) set to 0 V, adjust R_2 and R_x for a 25-kHz sine wave with minimum distortion at the output of the BPSK modulator.

3. Connect the carrier output from the BPSK modulator to both the recovered carrier input to the demodulator and to the external-trigger input of the oscilloscope.

4. Adjust the dc bias supply (V_C) to 5 V dc.

5. Sketch the modulator BPSK output waveform (V_{out}), and measure the demodulator output voltage (V_{out}).

6. In the modulator, increase the dc bias supply (V_C) to 7 V dc.

7. Sketch the modulator BPSK output waveform (V_{out}), and measure the demodulator output voltage (V_{out}).

8. Describe what effect changing V_C from 5 to 7 V dc has on the modulator output waveform and the demodulator output voltage.

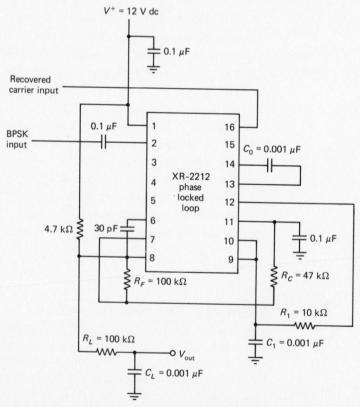

FIGURE 29-3   BPSK demodulator.

9. In the modulator, adjust the dc bias supply (V_C) to 6 V dc, and set the amplitude of the function generator output for a 4 Vp-p square wave with a frequency equal to 100 Hz.

10. Sketch the modulator binary input and demodulator binary output voltage waveforms.

11. Describe the relationship between the binary input and output waveforms sketched in step 10.

12. Vary the function generator output frequency, and describe what effect varying it has on the demodulator output waveform.

13. Disconnect the recovered carrier input signal from the demodulator, and describe what effect disconnecting it has on the demodulator output waveform.

SECTION C   Summary

Write a brief summary of the concepts presented in this experiment on BPSK demodulators. Include the following items:

1. The operation of a balanced demodulator (product detector).

2. The relationship between the dc bias supply voltage (V_C) and the modulator output phase.

3. The relationship between the binary input data, the output from the BPSK modulator, and the demodulator binary output signal.

EXPERIMENT 29 ANSWER SHEET

NAME: _____ CLASS: _____ DATE: _____

SECTION A

8.

Vertical sensitivity _____ V/cm

Time base _____ sec/cm

9. _____

_____

_____

SECTION B

5.

Vertical sensitivity _____ V/cm

Time base _____ sec/cm

Vertical sensitivity _____ V/cm

Time base _____ sec/cm

8. _____

_____

_____

10.

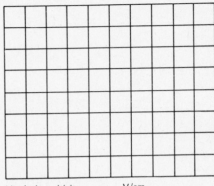

Vertical sensitivity _____ V/cm

Time base _____ sec/cm

Vertical sensitivity _____ V/cm

Time base _____ sec/cm

11. _____

_____

_____

12. _____

_____

_____

13. _____

_____

_____

BPSK TRANSMITTER RECEIVER DESIGN

REFERENCE TEXT: Electronic Communications Systems: Fundamentals through Advanced

1. Chapter 5, Phase-Locked Loop.
2. Chapter 5, Balanced Ring Modulator.
3. Chapter 8, PLL-FSK Demodulator.
4. Chapter 13, Binary Phase Shift Keying.
5. Chapter 13, BPSK Receiver.
6. Chapter 14, Synchronous Modems.

OBJECTIVES

1. To design, construct, test, and troubleshoot a BPSK transmitter to a given set of design specifications.
2. To design, construct, test, and troubleshoot a BPSK receiver to a given set of design specifications.
3. To assemble a complete BPSK transmitter/receiver system.

INTRODUCTION

In this experiment a complete operational BPSK transmitter and receiver system will be designed, constructed, and tested to a given set of specifications. The BPSK transmitter will be constructed using the XR-2206 monolithic function generator as the basic building block, and the BPSK receiver will be constructed using the XR-2212 monolithic phase-locked loop as the basic building block. Design equations and examples given in the manufacturer's specification sheets for the two integrated-circuit special-function chips will be used to determine the basic circuit configurations and to calculate specific component values. The simplified block diagram for the BPSK transmitter/receiver system that will be designed in this experiment is shown in Figure 30-1.

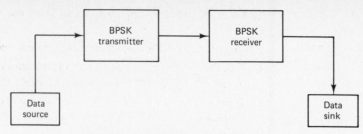

FIGURE 30-1   BPSK transmitter/receiver system.

MATERIALS REQUIRED

Equipment:

1 — protoboard

1 — dual dc power supply (+12 V dc and +6 V dc to −6 V dc)

1 — function generator (100 kHz)

1 — standard oscilloscope (10 MHz)

1 — assortment of test leads and hookup wire

Parts list:

1 — XR-2206 monolithic function generator

1 — XR-2212 monolithic phase-locked loop

1 — assortment of resistors and capacitors whose value depends on the particular design requirements

SECTION A   Design Specifications

In this section the design specifications for the BPSK transmitter/receiver system that will be designed in this experiment are given. The system will be designed to operate in the simplex (one-way only) mode. Therefore, only one BPSK transmitter and one BPSK receiver will be constructed and tested. In practice, BPSK is seldom used. Therefore, the BPSK system specifications for this experiment are purely hypothetical. However, the specifications are similar to those of the Bell System 212 modem. The specifications are given in Table 30-1.

TABLE 30-1   BPSK Modem Specifications

Center frequency:	1600 Hz
Operating mode:	simplex
Timing:	synchronous
Modulation:	BPSK
Bit rate:	600 bps

SECTION B   BPSK Transmitter Design

In this section the design equations given in the XR-2206 specification sheet are used to determine specific component values and operating parameters for the BPSK transmitter.

Procedure

1. Calculate the following system parameters: maximum fundamental input frequency, maximum bandwidth, maximum upper side frequency, minimum lower side frequency, and output spectrum.

2. Calculate the following component values and operating voltages: R_1, C_1, V^+, V_r, V_C.

3. Construct the BPSK modulator circuit designed in steps 1 and 2.

4. Test the BPSK modulator circuit constructed in step 3 to determine if the modulator is operating within the design specifications given in Section A.

5. Troubleshoot the BPSK modulator as required, and make any design changes, adaptions, or additions that are necessary for the transmitter to operate within the given specifications.

SECTION C   BPSK Receiver Design

In this section the design equations given in the XR-2212 specification sheet are used to determine specific component values and operating parameters for the BPSK receiver.

Procedure

1. Calculate the following receiver component values and system parameters.

   Component values: R_0, C_0, R_1, R_C, and R_F. System parameters: τ, Γ, loop-tracking bandwidth, K_0, K_ϕ, and K_T.

2. Construct the BPSK receiver circuit designed in step 1.

3. Test the BPSK receiver circuit constructed in step 2 to determine if the receiver is operating within the design specifications given in Section A.

4. Troubleshoot the BPSK receiver as required, and make any design changes, adaptions, or additions that are necessary for the receiver to operate within specifications.

SECTION D   Circuit Implementation

In this section the BPSK transmitter and receiver circuits designed, constructed, and tested in Sections B and C are combined and tested as a complete system. Make any adjustments or changes necessary for the system to operate within the given specifications.

SECTION E   Summary

Write a complete laboratory report for the BPSK transmitter/receiver system designed and tested in Sections B, C, and D. Include the following items:

1. Complete schematic diagrams of the transmitter and the receiver.

2. An introduction including a statement of the problem.

3. All necessary calculations.

4. All pertinent data.

5. A list of problems encountered and how they were rectified.

6. A conclusion.

EXPERIMENT 30 ANSWER SHEET

NAME: _____ CLASS: _____ DATE: _____

SECTION B

1. Maximum fundamental input frequency = _____
   Maximum bandwidth = _____
   Maximum upper side frequency = _____
   Minimum lower side frequency = _____
   Output spectrum (below)

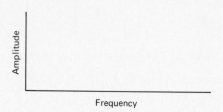

2. R_1 = _____ V_r = _____
   C_1 = _____ V_C = _____
   V^+ = _____

SECTION C

1. R_0 = _____ R_C = _____
   C_0 = _____ R_F = _____
   R_1 = _____
   τ = _____
   Γ = _____
   Loop-tracking bandwidth = _____
   K_O = _____
   K_ϕ = _____
   K_T = _____

RS-232C
INTERFACE DESIGN

REFERENCE TEXT: Electronic Communications Systems: Fundamentals through Advanced

1. Chapter 14, Serial Interfaces.
2. Chapter 14, RS-232C Interface.

OBJECTIVES

1. To observe the operation of an RS-232C interface.
2. To design an RS-232C interface to meet a given set of design specifications.
3. To observe the transmission characteristics of RS-232C signals.

INTRODUCTION

A special interface is placed between a line-control unit (LCU) and a data modem to ensure an orderly flow of data. This interface coordinates the flow of data, control signals, and timing information between the LCU and the modem. The Electronics Industries Association (EIA) standardized the RS-232C serial interface to facilitate interconnecting digital terminal equipment (DTE) and digital communications equipment (DCE) that are manufactured by different companies. The RS-232C interface specifications identify the mechanical, electrical, and functional description for the interface between DTEs and DCEs. The RS-232C interface is simply a cable and two connectors. However, the standard also specifies limitations on the DCE and DTE input and output impedances and the voltage levels and rise and fall times of the signals that propagate across the interface.

The RS-232C interface is a 25-wire cable with a DB25P/DB25S compatible connector. The pins on the RS-232C interface cable are functionally categorized as either ground, data, control, or timing. All the pins are unidirectional (signals propagate only from the DTE to the DCE or vice versa). Table 31-1 lists the 25 pins of the RS-232C interface, their designations, and the direction of signal propagation. In this experiment only the electrical and functional specifications for the first eight pins on the RS-232C interface are examined.

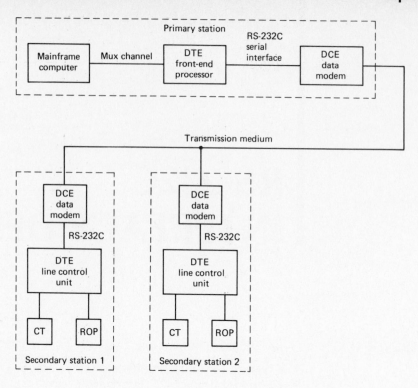

MATERIALS REQUIRED

Equipment:

1 — protoboard

1 — dual dc power supply (+ 15 V dc to − 15 V dc)

1 — function generator (10 kHz)

1 — standard oscilloscope (10 mHz)

1 — assortment of test leads and hookup wire

Parts list:

1 — assortment of resistors, capacitors, and integrated circuits, depending on the individual circuit implementation.

SECTION A   Functional Description

In this section the functional descriptions for the first eight pins on the RS-232C interface are given. The block diagram for the circuit used in this experiment is shown in Figure 31-1.

Pin 1—protective ground. This pin is the frame ground and is used for protection against electrical shock. Pin 1 should be connected to the third-wire ground of the electrical system at one end of the cable (either at the DTE or the DCE, but not at both ends).

Pin 2—transmit data (TD). Serial data on the primary channel from the DTE to the DCE are transmitted on this pin. TD is enabled by an active condition on the CS pin (pin 5).

Pin 3—received data (RD). Serial data on the primary communications channel from the

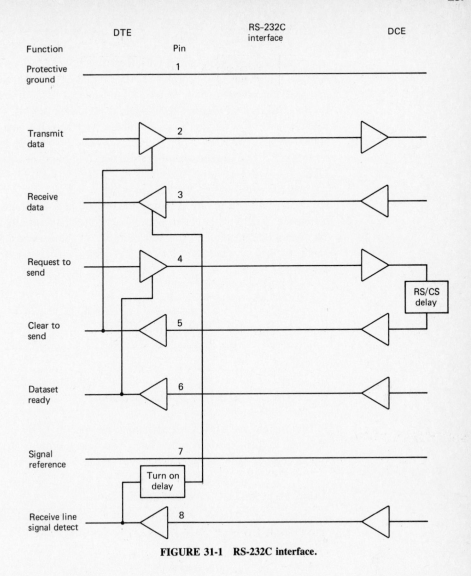

FIGURE 31-1   RS-232C interface.

DCE to the DTE are transferred on this pin. RD is enabled by an active condition on the RLSD pin (pin 8).

Pin 4—request to send (RS). The DTE bids for the primary communications channel from the DCE on this pin. An active condition on RS turns on the modem's carrier. The analog carrier "trains" the receiver and initializes the communications channel. RS cannot go active unless DSR (pin 6) is active.

Pin 5—clear to send (CS). This signal is a handshake from the DCE to the DTE in response to an active condition on request to send. CS enables the TD pin.

Pin 6—data set ready (DSR). On this pin the DCE indicates the availability of the communications channel. DSR is active as long as the DCE is connected to the communications channel.

Pin 7—signal ground. This pin is the signal reference for all the data, control, and timing pins. Usually, this pin is strapped to frame ground (pin 1).

Pin 8—receive-line-signal detect (RLSD). The DCE uses this pin to signal the DTE when the DCE is receiving an analog carrier on the primary data channel. RLSD enables the RD pin.

SECTION B   Electrical Specifications

In this section the electrical specifications for the RS-232C interface are described. The RS-232C interface specifications impose limitations on the voltage levels that the DTE and DCE can output onto or receive from the cable. In both the DTE and DCE, there are leveler circuits that convert their internal logic levels to RS-232C levels. A leveler is called a driver if it outputs a signal voltage onto the cable and a terminator if it accepts a signal voltage from the cable. Table 31-2 lists the voltage limits for both drivers and terminators. Note that the data lines use negative logic and the control lines use positive logic.

A driver can output any voltage between $+5$ and $+15$ or -5 and -15 V dc, and a terminator will accept any voltage between $+3$ and $+25$ and -3 and -25 V dc. The difference in the voltage levels between a driver and a terminator is called noise margin. The noise margin reduces the susceptibility of the interface to noise transients on the cable. Typical voltages used for data and control signals are $+/-7$ V dc and $+/-10$ V dc.

Stop bit (1, 1.5, 2)	Parity bit		Data bits (5–7)							Start bit
1	1	1/0	b_6 MSB	b_5	b_4	b_3	b_2	b_1	b_0	0 LSB

SECTION C   Design Specifications

In this section the design specifications for the RS-232C interface circuit that will be designed in this experiment are given. The binary signal source can be simulated with a function generator. The interface specifications are given in Table 31-3.

TABLE 31-3   RS-232C Interface Specifications

		Driver output voltage (Vdc)	Terminator input voltage (Vdc)
Date pins	logic 1	-5 to -15	-3 to -25
(2 and 3)	logic 0	$+5$ to $+15$	$+3$ to $+25$
Control pins	enable	$+5$ to $+15$	$+3$ to $+25$
(4, 5, 6 and 8)	disable	-5 to -15	-3 to -25
RS/CS delay		40 ms	
RLSD turn-on delay		10 ms	
RLSD turn-off delay		0 ms	
Bit rate		1200 bps	
Driver input and terminator output voltages:			logic 1 = +5 Vdc logic 0 = 0 Vdc

SECTION D   Circuit Design and Implementation

In this section an RS-232C interface will be designed to meet the specifications given in Table 31-3. Only the first eight pins will be included. The block diagram is shown in Figure 31-1.

Procedure

1. Design the data levelers for the RS-232C interface circuit shown in Figure 31-1 (drivers and terminators) to the specifications given in Table 31-3.

2. Design the control levelers for the RS-232C interface circuit shown in Figure 31-1 (drivers and terminators) to the specifications given in Table 31-3.

3. Construct the data and control levelers designed in steps 1 and 2 for the TD, RD, RS, CS, DSR, adn RLSD circuits shown in Figure 31-2.

4. Test the levelers constructed in step 3 to determine if they are operating within the specifications given in Table 31-3.

5. Troubleshoot the levelers tested in step 4, and make any design changes, adaptions, or additions that are necessary for them to operate within the given specifications.

6. Design the 40 ms RS/CS delay circuit to the specifications given in Table 31-3.

7. Design the RLSD turn-on delay circuit to the specifications given in Table 31-3.

8. Test the delay circuits designed in steps 6 and 7 to determine if they are operating within the specifications given in Table 31-3.

9. Troubleshoot the delay circuits tested in step 8, and make any design changes, adaptions, or additions that are necessary for them to operate within the given specifications.

10. Connect the appropriate drivers and terminators, and test the overall operation of the interface for an input bit rate $F_b = 1200$ bps.

11. Troubleshoot the system tested in step 10, and make any design changes, adaptions, or additions that are necessary for it to operate within the given specifications.

SECTION E   Summary

Write a complete laboratory report for the RS-232C interface circuit designed, constructed, and tested in Section D. Include the following items:

1. Complete schematic diagrams for the drivers, terminators, and delay circuits.

2. An introduction, including a statement of the problem.

3. All pertinent data.

4. A list of problems encountered and how they were rectified.

5. A conclusion.

PARITY GENERATOR/CHECKER

REFERENCE TEXT: Electronic Communications Systems: Fundamentals through Advanced

1. Chapter 14, Parity.

OBJECTIVES

1. To observe the operation of sequential and combinational parity generators.
2. To observe the operation of sequential and combinational parity checkers.

INTRODUCTION

Parity is probably the simplest error detection scheme used for data communications systems. With parity, a single bit (called the parity bit) is added to each character to force the total number of 1s in the character, including the parity bit, to be either an odd number (odd parity) or an even number (even parity). The definition of *parity* is equivalence or equality. A logic gate that will determine when all its inputs are equal is the XOR gate. With an XOR gate, if all the inputs are equal (either all 0s or all 1s), the output is a 0. If all inputs are not equal, the output is a 1. There are two circuit configurations using XOR gates that are commonly used to generate a parity bit: sequential (series) and combinational (parallel) circuits. Both types of parity generators can also be used for a parity checker. A parity checker examines a received character and determines if a parity error has occurred. In this experiment, the operation of both sequential and parallel parity generators and parity checkers are examined.

MATERIALS REQUIRED

Equipment:

1 — protoboard
1 — dc power supply (+ 12 V dc)
1 — assortment of test leads and hookup wire

Parts list:

1 — CD4070B quad two-input XOR gate
5 — SPDT switches
5 — 1k-ohm resistors

SECTION A   The Sequential Parity Generator and Parity Checker

In this section the operation of a sequential parity generator and parity checker is examined. The schematic diagram for the parity generator/checker circuit used in this section is shown in Figure 32-1. With sequential parity generation, b_0 is XORed with b_1, the result is XORed with b_2, and so on. The result of the last XOR operation is compared (XORed) with a bias bit. If even parity is desired, the bias bit is made a logic 0. If odd parity is desired, the bias bit is made a logic 1. The output of the circuit is the parity bit, which is appended to the character code prior to transmission. The circuit shown in Figure 32-1 can also be used for a sequential parity checker. The operation is the same except that the logic condition of the final comparison indicates whether a parity error has occurred or not. For odd parity, a logic 1 indicates that an error has occurred and a logic 0 indicates that no error has occurred. For even parity, a logic 0 indicates that an error has occurred and a logic 1 indicates that no error has occurred.

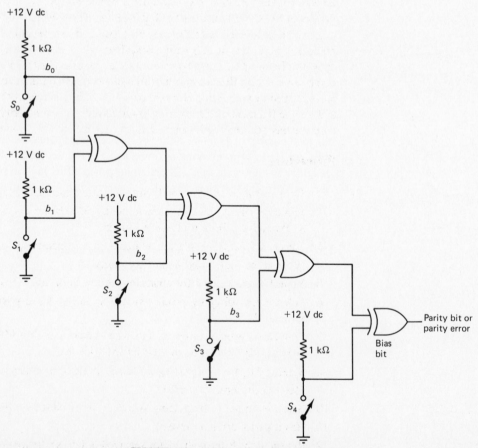

FIGURE 32-1   Serial parity generator/checker.

Procedure

1. Construct the sequential parity generator/checker circuit shown in Figure 32-1.
2. Simulate a 4-bit transmit character with a 0000 binary sequence by placing switches S_0 through S_3 in their closed positions.
3. Select odd parity by placing switch S_4 in its open position.
4. What is the logic condition of the parity bit?
5. Repeat steps 3 and 4 for character sequences of 0001 through 1111.
6. Select even parity by placing switch S_4 in its closed position, and repeat steps 2 through 5.
7. Simulate a 4-bit even-parity received character with a 0000 sequence by placing switches S_0 through S_3 in their closed positions.
8. Select odd parity by placing switch S_4 in its open position.
9. Has a parity error occurred?
10. Select even parity by placing switch S_4 in its closed position.
11. Has a parity error occurred?
12. Repeat steps 8 through 11 for received binary sequences of 0001 through 1111.

SECTION B   The Combinational Parity Generator and Parity Checker

In this section the operation of a combinational parity generator and parity checker is examined. The schematic diagram for the parity generator/checker circuit used in this section is shown in Figure 32-2. With parallel parity generation, comparisons are made in layers or levels. Pairs of bits (b_0 and b_1, b_2 and b_3, etc.) are XORed. The results of the first-level XOR gates are then XORed together. The process continues until only one bit is left, which is XORed with the bias bit. Again, if even parity is desired, the bias bit is made a logic 0, and if odd parity is desired, the bias bit is made a logic 1. The circuit shown in Figure 32-2 can also be used for a combinational parity checker. The operation is the same except that the result of the final comparison indicates whether a parity error has occurred or not. Again, for odd parity, a logic 1 indicates that an error has occurred and a logic 0 indicates that no error has occurred. For even parity, a logic 0 indicates that no error has occurred and a logic 1 indicates that an error has occurred.

Procedure

1. Construct the combinational parity generator/checker circuit shown in Figure 32-2.
2. Simulate a 4-bit transmit character with a 0000 binary sequence by placing switches S_0 through S_3 in their closed positions.
3. Select odd parity by placing switch S_4 in its open position.
4. What is the logic condition of the parity bit?
5. Repeat steps 3 and 4 for character sequences of 0001 through 1111.
6. Select even parity by placing switch S_4 in its closed position and repeat steps 2 through 5.
7. Simulate a 4-bit even-parity received character with a 0000 sequence by placing switches S_0 through S_3 in their closed positions.
8. Select odd parity by placing switch S_4 in its open position.
9. Has a parity error occurred?
10. Select even parity by placing switch S_4 in its closed position.
11. Has a parity error occurred?
12. Repeat steps 8 through 11 for received binary sequences of 0001 through 1111.

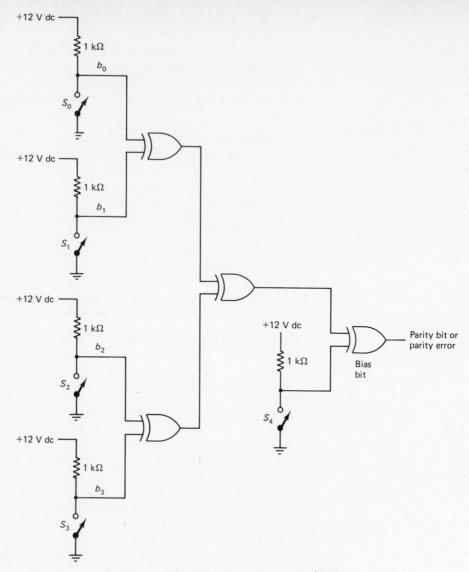

FIGURE 32-2   Combinational parity generator/checker.

SECTION C   Summary

Write a brief summary of the concepts presented in this experiment on parity generation and parity checking. Include the following items:

1. The concept of odd and even parity.
2. The operation of an XOR gate.
3. The operation of a sequential parity generator and checker.
4. The operation of a combinational parity generator and checker.
5. The significance of a parity error.

EXPERIMENT 32 ANSWER SHEET

NAME: _____ CLASS: _____ DATE: _____

SECTION A

4. Logic condition (0000) = _____

5.　　　Parity

			Parity
0000 _____		1000 _____	
0001 _____		1001 _____	
0010 _____		1010 _____	
0011 _____		1011 _____	
0100 _____		1100 _____	
0110 _____		1110 _____	
0111 _____		1111 _____	

6.　　　Parity

			Parity
0000 _____		1000 _____	
0001 _____		1001 _____	
0010 _____		1010 _____	
0011 _____		1011 _____	
0100 _____		1100 _____	
0101 _____		1101 _____	
0110 _____		1110 _____	
0111 _____		1111 _____	

9. _____

11. _____

12.　　　Parity error

			Parity error
0000 _____		1000 _____	
0001 _____		1001 _____	
0010 _____		1010 _____	
0011 _____		1011 _____	
0100 _____		1100 _____	
0101 _____		1101 _____	
0110 _____		1110 _____	
0111 _____		1111 _____	

SECTION B

4. Logic condition (0000) = _____

5.　　　Parity

			Parity
0000 _____		1000 _____	
0001 _____		1001 _____	
0010 _____		1010 _____	
0011 _____		1011 _____	
0100 _____		1100 _____	
0101 _____		1101 _____	
0110 _____		1110 _____	
0111 _____		1111 _____	

6.   Parity   Parity

    0000 _____   1000 _____

    0001 _____   1001 _____

    0010 _____   1010 _____

    0011 _____   1011 _____

    0100 _____   1100 _____

    0101 _____   1101 _____

    0110 _____   1110 _____

    0111 _____   1111 _____

9. _____

11. _____

12.   Parity error   Parity error

    0000 _____   1000 _____

    0001 _____   1001 _____

    0010 _____   1010 _____

    0011 _____   1011 _____

    0100 _____   1100 _____

    0101 _____   1101 _____

    0110 _____   1110 _____

    0111 _____   1111 _____

UNIVERSAL ASYNCHRONOUS RECEIVER/ TRANSMITTER

REFERENCE TEXT: Electronic Communications Systems: Fundamentals through Advanced

1. Chapter 14, Parity.
2. Chapter 14, Asynchronous Data Format.
3. Chapter 14, Universal Asynchronous Receiver/Transmitter.

OBJECTIVES

1. To observe the operation of a universal asynchronous receiver/transmitter.
2. To observe the operation of a special-purpose large scale integration universal asynchronous receiver/transmitter.
3. To observe the transmission of asynchronous data.

INTRODUCTION

A universal asynchronous receiver/transmitter (UART) is used for asynchronous transmission of data between a DTE and a DCE. Asynchronous transmission means that an asynchronous data format is used and there is no clocking information transferred between the DTE and the DCE. There are three primary functions of the UART:

1. To perform serial-to-parallel and parallel-to-serial conversion of data.
2. To perform error detection by inserting and checking parity bits.
3. To insert, detect, and delete start and stop bits.

In this experiment, the Intersil IM6402 large scale integration UART is used. The IM6402 is a CMOS/LSI UART for interfacing computers or microprocessors to asynchronous serial data channels. The UART is functionally divided into two sections: the transmitter and the receiver. The transmitter converts parallel data into serial form and automatically adds start, parity, and stop bits. The receiver converts serial start, data, parity, and stop bits to parallel data and verifies proper code, parity, start, and stop bit transmission. The data-word length can be 5, 6, 7, or 8 bits. Parity may be odd or even, and parity checking and generation can be inhibited. The stop bits may be 1 or 2 (or 1.5 when transmitting a 5-bit code). The functional block diagram for the IM6402 UART is shown in Figure 33-1a, and the pin configuration is shown in Figure 33-1b.

MATERIALS REQUIRED

Equipment:

1 — protoboard
1 — dc power supply (+8 V dc)
2 — function generators (50 kHz each)
1 — standard oscilloscope (10 MHz)
1 — assortment of test leads and hookup wire

Parts list:

1 — IM6402 LSI UART
1 — CD4069B hex inverter
11 — 1k-ohm resistors*
11 — light emitting diodes (LEDs)*
13 — SPDT DIP switches

SECTION A   The UART Control Word Register

In this section the operation of the UART control word register is examined. Prior to transferring data in either direction, a control word must be programmed into the UART control register to indicate the nature of the data (such as the number of data bits), whether parity is used (and if so, whether it is even or odd), and the number of stop bits. Essentially, the start bit is the only bit that is not optional; there is always only one start bit and it must be a logic 0. In the UART transmitter, the control word is used to set up the data-, parity-, and stop-bit steering logic circuit. In the UART receiver, the control register is used for data-, parity-, and stop-bit detection. Table 33-1 shows how to program the control word for the IM6402 UART.

The two character-length-select inputs (CLS_1 and CLS_2) determine the number of data bits in the transmitted and received word. The parity inhibit input (PI) determines if a parity bit is included in the transmitted word and checked in the received word. The even-parity-enable (EPE) input determines whether even or odd parity is used. The stop-bit-select input (SBS) determines how many stop bits are transmitted at the end of each character and checked in the receiver.

SECTION B   UART Transmitter

In this section the operation of the UART transmitter is examined. The schematic diagram for the UART transmitter/receiver used in this section is shown in Figure 33-2a. The UART transmitter accepts parallel data, formats it, and transmits it in asynchronous serial form. Figure 33-2b shows the asynchronous serial data format used with the IM6402 UART, and the transmitter timing diagram is shown in Figure 33-2c. The parallel transmit

---

*The eleven 1k ohm resistors and light emitting diodes are necessary only if a visual display of the received data and status bits is desired.

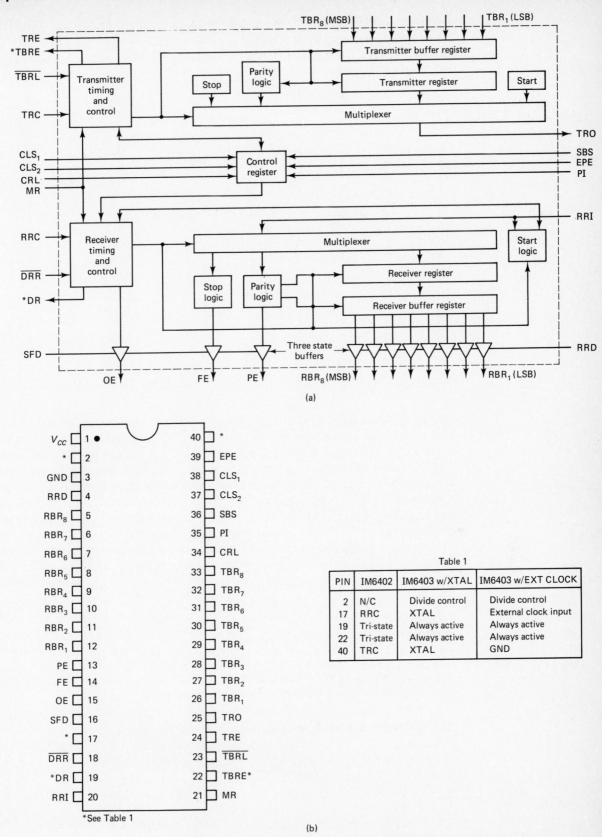

Table 1

PIN	IM6402	IM6403 w/XTAL	IM6403 w/EXT CLOCK
2	N/C	Divide control	Divide control
17	RRC	XTAL	External clock input
19	Tri-state	Always active	Always active
22	Tri-state	Always active	Always active
40	TRC	XTAL	GND

*See Table 1

(b)

FIGURE 33-1   IM6402 UART. (a) Functional block diagram. (b) Pin configuration.

TABLE 33-1   IM6402 Control-word programming

CLS^2	CLS^1	Pl	EPE	SBS	Data bits	Parity bit	Stop bits(S)
		Control word					
L	L	L	L	L	5	Odd	1
L	L	L	L	H	5	Odd	1.5
L	L	L	H	L	5	Even	1
L	L	L	H	H	5	Even	1.5
L	L	H	X	L	5	Disabled	1
L	L	H	X	H	5	Disabled	1.5
L	H	L	L	L	6	Odd	1
L	H	L	L	H	6	Odd	2
L	H	L	H	L	6	Even	1
L	H	L	H	H	6	Even	2
L	H	H	X	L	6	Disabled	1
L	H	H	X	H	6	Disabled	2
H	L	L	L	L	7	Odd	1
H	L	L	L	H	7	Odd	2
H	L	L	H	L	7	Even	1
H	L	L	H	H	7	Even	2
H	L	H	X	L	7	Disabled	1
H	L	H	X	H	7	Disabled	2
H	H	L	L	L	8	Odd	1
H	H	L	L	H	8	Odd	2
H	H	L	H	L	8	Even	1
H	H	L	H	H	8	Even	2
H	H	H	X	L	8	Disabled	1
H	H	H	X	H	8	Disabled	2

data word is determined by the parallel binary input-word switches and loaded into the transmitter buffer register from parallel inputs TBR_1 TO TBR_8 on the negative transition of the $\overline{TBRL}$ (transmitter buffer register load) input. The rising edge of $\overline{TBRL}$ clears TBRE (transmit buffer register empty). Zero to one clock cycles later, data is transferred to the transmitter register. TRE (transmit register empty) is cleared, and serial transmission begins. TBRE is reset to a logic high. Output data are clocked by TRC (transmit register clock), which is 16 times the data rate. A second pulse on $\overline{TBRL}$ loads data into the transmitter buffer register. Data transfer to the transmitter register is delayed until transmission of the current character is complete. Data is automatically transferred to the transmitter register, and transmission of that character begins. If $\overline{TBRL}$ is continuously pulsed without the parallel data input word changing, the same data word will be transmitted repeatedly.

Procedure

1. Construct the UART transmitter-receiver circuit shown in Figure 33-2a (only the transmit portion of the circuit is used in this section).

2. Program the control word register for 8 data bits, odd parity, and 2 stop bits by placing the following switches in the indicated positions:

$$CLS_1 \quad CLS_2 \quad PI \quad EPE \quad SBS$$
$$1 \qquad 1 \qquad 0 \qquad 0 \qquad 1$$

3. Load the control word into the control register by switching the CRL input (pin 34) to a high.

4. Set the amplitude of the transmit clock function generator output voltage to an 8 V-p–p square wave with a frequency $F_{trc} = 32$ kHz ($V_{max} = +8$ V and $V_{min} = 0$ V).

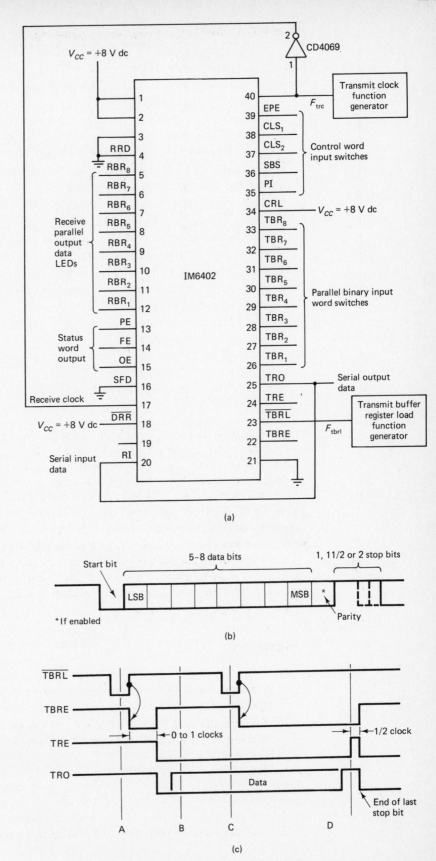

FIGURE 33-2   UART Transmitter/receiver. (a) Schematic diagram. (b) A-Synchronous data format. (c) Timing diagram.

5. Set the amplitude of the transmit buffer register load ($\overline{\text{TBRL}}$) function generator output voltage to an 8 Vp–p square wave with a frequency $F_{\text{tbr1}} = 1$ kHz ($V_{\text{max}} = +8$ V and $V_{\text{min}} = 0$ V).

6. Set the input data switches for a parallel binary input word of 10101010 (TBR$_1$ to TBR$_8$).

7. Observe the waveform for the transmitted serial data word at the transmit register output (TRO) on pin 25 (it may be necessary to adjust the $\overline{\text{TBRL}}$ function generator output frequency slightly to observe a stable output waveform).

8. Sketch the waveform observed in step 6, and identify each of the following bits: start bit, data bits, parity bit, and stop bits.

9. Change the parallel binary input word to 10010010, and repeat steps 7 and 8.

10. Change the parity odd/even (EPE) switch to a low, and describe what effect changing it has on the output waveform (it may be necessary to adjust the $\overline{\text{TBRL}}$ function-generator output frequency slightly to observe a stable waveform).

11. Change the stop bit select (SBS) switch to a low, and describe what effect changing it has on the output waveform (it may be necessary to change the $\overline{\text{TBRL}}$ function generator output frequency slightly to observe a stable waveform).

12. Change the character length select (CLS$_1$ and CLS$_2$) switches to 00, respectively, and describe what effect changing them has on the output waveform (it may be necessary to change the $\overline{\text{TBRL}}$ function generator output frequency slightly to observe a stable waveform).

13. Change the parity inhibit (PI) switch to a high, and describe what effect changing it has on the output waveform (it may be necessary to change the $\overline{\text{TBRL}}$ function generator output frequency slightly to observe a stable output waveform).

14. Determine the control word input switch positions necessary to transmit a character with the following characteristics: 1 start bit, 5 data bits, odd parity, and 1.5 stop bits.

15. Place the control word input switches into the positions determined in step 14.

16. Verify the proper transmission by observing the TRO output.

17. Restore the control-word input switches to the conditions shown in step 2.

18. Leave this circuit assembled, as it is needed in Section C.

SECTION C The UART Receiver

In this section the operation of a UART receiver is examined. The schematic diagram for the UART transmitter/receiver circuit used in this section is shown in Figure 33-2a. Serial data is received at the RI input. When no data is being received, the RI input must remain high to simulate idle line 1s. The receiver timing diagram is shown in Figure 33-3. Data are clocked into the receiver after detection of a valid start bit. A low level on DRR (data-receive reset) clears the DR (data ready) line. During the first stop bit, data are transferred from the receiver register to RBR (receive buffer register). If the word is less than 8 bits long (as determined by the control register), the unused most significant bits are held low. The output character is right justified to the least significant bit (RBR$_1$). The receiver status register outputs information concerning the integrity of the received data. The data word length, parity, and stop-bit length are determined by the control register. There is only one control register; therefore, the transmitter and receiver word formats must be the same. A logic high on status output OE indicates that a receiver overrun has occurred, which means that DR was not cleared before the present character was transferred to RBR. A logic high on status output PE indicates that a parity error has occurred. A logic

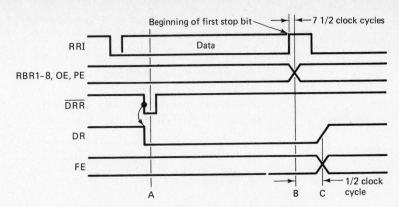

FIGURE 33-3 IM6402 Receiver Timing Diagram.

high on status output FE indicates that a framing error has occurred, which means that either no, or an improper number of, stop bits were received.

Procedure

1. The UART transmitter/receiver constructed in step 1 of Section B is also used for this section.
2. Verify the proper operation of the UART transmitter, and verify that the control word is as shown in step 2 of Section B.
3. Connect LEDs to the receive parallel output data pins (RBR_1 to RBR_8) and to the status word outputs (pins 13 to 15).
4. Set the input-data switches for a parallel binary input word of 11001100 (TBR_1 to TBR_8).
5. Sketch the waveform for the transmitted serial data word at the transmit register output (TRO).
6. Compare the received binary word displayed on the receive parallel output LEDs to the transmitted binary word. Are they the same?
7. Change the input data switches for a transmitted serial data word of 00110011, and repeat steps 5 and 6.
8. What are the conditions for the following status word outputs: PE, FE, and OE? Are these the correct conditions?
9. Change the data word length to 5 bits by changing the CLS_1 and CLS_2 inputs to 00, respectively.
10. Sketch the transmit waveform, and label the start, data, parity, and stop bits.
11. Describe the parallel received data word displayed on the TBR_1 to TBR_8 output LEDs.
12. Disconnect the UART transmitter output from the receiver input.
13. Simulate a random received serial data input word by connecting the transmit buffer register-load function generator output (F_{tbr1}) to the UART receiver input (DI).
14. Adjust the frequency of the transmit buffer register load function generator until a stable binary sequence is displayed on the receive parallel output LEDs.
15. What are the conditions for the following status word outputs: PE, FE, and OE?
16. What do the indications observed in step 15 indicate?
17. Change the frequency of the transmit buffer register load function generator until a different stable binary sequence is displayed on the receive parallel-output LEDs.
18. Repeat steps 15 and 16.

SECTION D   Summary

Write a brief summary of the concepts presented in this experiment on UART transceivers. Include the following items:

1. The basic operation of a UART transmitter.
2. The basic operation of a UART receiver.
3. The purpose of the control and status registers.
4. The concepts of parity and asynchronous data transmission.
5. The meaning of the following status bits: PE, FE, and OE.

EXPERIMENT 33 ANSWER SHEET

NAME: _____ CLASS: _____ DATE: _____

SECTION B

8.

Vertical sensitivity _____ V/cm

Time base _____ sec/cm

9.

Vertical sensitivity _____ V/cm

Time base _____ sec/cm

10. _____
_____
_____

11. _____
_____
_____

12. _____
_____
_____

13. _____
_____
_____

14. CLS_1 = __ CLS_2 = __ PI = __ EPE = __ SBS = __

SECTION C

5.

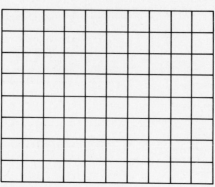

Vertical sensitivity _____ V/cm

Time base _____ sec/cm

6. _____

_____

_____

7.

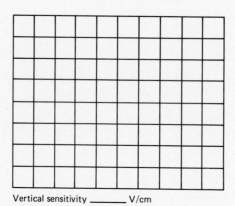

Vertical sensitivity _____ V/cm

Time base _____ sec/cm

_____

_____

_____

8. PE = __ FE = __ OE = __

10.

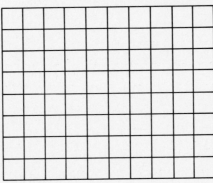

Vertical sensitivity _____ V/cm

Time base _____ sec/cm

11. _____

_____

_____

15. PE = __ FE = __ OE = __

16. _____

_____

_____

18. PE = __ FE = __ OE = __

_____

_____

_____

Experiment 34

DIGITAL FILTERS

REFERENCE TEXT: Specification Sheet for the MF10 Universal Monolithic Dual Switched Capacitor Filter

OBJECTIVES

1. To observe the operation of a linear integrated-circuit digital filter.
2. To observe the operation of a switched capacitor filter.
3. To observe the frequency response characteristics of digital lowpass, bandpass, and bandstop filters.

INTRODUCTION

When designing modern electronic communications systems, a great deal of emphasis is placed on miniaturization. Consequently, large scale integrated circuits have been developed for many of the circuits commonly used in electronic communications systems, such as filters. Integrated circuits offer more compact packages, less power consumption, lower cost, simplicity in design, improved noise immunity, and better overall performance than their discrete counterparts. In this experiment the operation of the MF10 universal monolithic dual switched capacitor filter is examined. The MF10 is a switched capacitor (sampled-data) filter that comes in a 20-pin dual-in-line package (DIP) which contains two independent general purpose CMOS active filter building blocks. Each block, together with an external clock and three or four external resistors, can produce various second-order (single-pole) filter functions (i.e., lowpass, highpass, bandpass, and bandstop). Each building block has three outputs. One of the outputs can be configured to perform either an allpass (AP), highpass (HP), or bandstop (BS) function. The two remaining outputs perform lowpass (LP) and bandpass (BP) functions. There are six modes available with the MF10, and some of the modes have more than one configuration. The block diagram for the MF10 universal switched-capacitor filter is shown in Figure 34-1a, and the DIP pin layout is shown in Figure 34-1b.

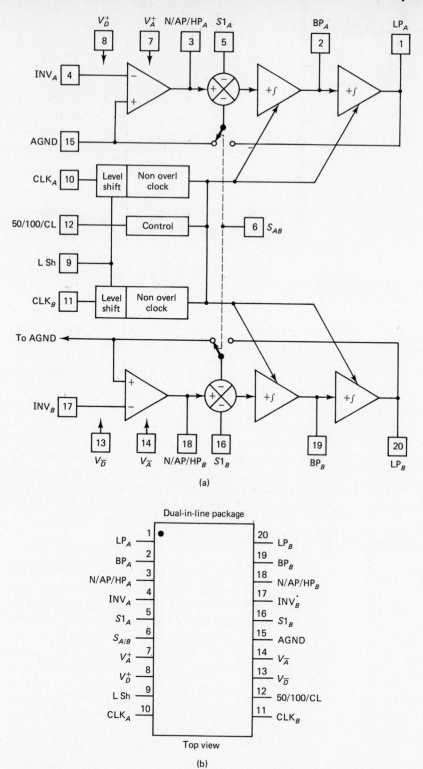

(a)

(b)

FIGURE 34-1 MF10 monolithic switched capacitor digital filter. (a) Block diagram. (b) DIP pin out.

MATERIALS REQUIRED

Equipment:

1 — protoboard

1 — dual dc power supply ($+5$ V dc and -5 V dc)

1 — low-frequency signal generator (10 kHz)

1 — medium-frequency clock source (200 kHz)

1 — standard oscilloscope (10 MHz)

1 — assortment of test leads and hookup wire

Parts List

1 — MF10 universal switched capacitor filter

2 — 1k-ohm reesistors

1 — 4.7k-ohm resistor

4 — 10k-ohm resistors

SECTION A   Mode 1 Bandstop, Bandpass, and Lowpass Operation — Second-Order

In this section the frequency-response characteristics of the MF10 switched-capacitor filter operating in mode 1 are examined. In mode 1 the MF10 can simultaneously perform bandpass, bandstop, and lowpass functions where the center frequency of the bandpass and notch filters and the break frequency for the lowpass filter are equal. The schematic diagram for the multipurpose digital filter circuit used in this section is shown in Figure 34-2a. The electrical model for mode 1 operation is shown in Figure 34-2b. External resistors R_{1a}, R_{2a}, and R_{3a} determine the gain, Q factor, and bandpass characteristics for the various filter configurations. The external clock determines the center frequency for the bandpass and bandstop filters and the break frequency for the lowpass filter.

Procedure

1. Construct the multipurpose single-pole filter circuit shown in Figure 34-2a (jumper J_1 disconnected).

2. Calculate the center frequency for the bandpass and notch filters and the break frequency for the lowpass filter using the following formula.

$$F_o = \frac{F_c}{100}$$

where F_o = filter center frequency
                F_c = external-clock frequency

3. Calculate the Q factor for the filter using the following formula.

$$Q = \frac{R_3}{R_2}$$

where Q = quality factor

4. Calculate the bandwidth of the bandpass and bandstop filters using the following formula.

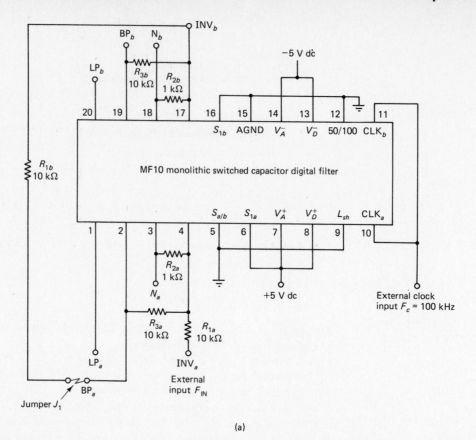

(a)

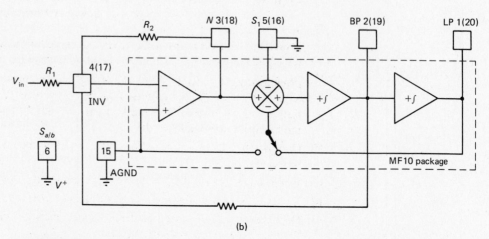

(b)

FIGURE 34-2 Bandstop, bandpass, and lowpass operation - 2nd order. (a) Schematic diagram. (b) Mode 1 model.

$$B = \frac{F_o}{Q}$$

where
B = bandwidth
F_o = filter center frequency
Q = Q factor

5. Calculate the gain of the lowpass filter using the following formula.

$$H_{olp} = \frac{-R_2}{R_1}$$

where H_{olp} = lowpass filter gain as F_{in} approaches 0 Hz

6. Calculate the gain of the bandpass filter using the following formula.

$$H_{obp} = \frac{-R_3}{R_1}$$

where H_{obp} = bandpass filter gain when F_{in} equals F_o

7. Calculate the gain of the bandstop filter using the following formula.

$$H_{on} = \frac{-R_2}{R_1}$$

where H_{on} = bandstop filter gain when F_{in} approaches 0 Hz or when F_{in} approaches the external-clock frequency divided by 2

8. Sketch the gain-versus-frequency-response curve for the bandpass (BP_a), bandstop (N_a), and lowpass (LP_a) filter outputs on semilog graph paper.
9. From the response curves sketched in step 8, determine the bandpass, bandstop, and lowpass filter gains, the bandpass filter center frequency, the bandstop filter center frequency, the lowpass filter break frequency, and the bandwidth and Q factor for the bandpass and bandstop filters.
10. Compare the results measured in steps 8 and 9 to the values calculated in steps 2 through 7.
11. Increase the clock frequency to 200 kHz, and repeat step 8.
12. Decrease the clock frequency to 50 kHz, and repeat step 8.
13. Describe what effect changing the clock frequency has on the response curves for the three filter configurations.
14. Reset the clock frequency to 100 kHz, replace R_{2a} with a 4.7k ohm resistor, and repeat step 8.
15. Describe what effect increasing the resistance of R_{2a} has on the response curves for the three filter configurations.

SECTION B   Mode 1 Bandstop and Bandpass Operation— Fourth-Order

In this section the frequency-response characteristics for two cascaded MF10 switched capacitor filter building blocks operating in mode 1 are examined. The MF10 is ideally suited for cascading filter blocks to perform higher than second-order filter functions. The schematic diagram for the fourth-order multipurpose digital filter circuit used in this section is also shown in Figure 34-2. The design equations for determining the center frequency for the bandpass and bandstop filters for fourth-order operation are identical to those used for second-order operation, and the overall filter gains are simply the product of the gains of the individual building blocks. The component values used in this section are identical to those used in Section A. Therefore, the center frequencies are the same and need not be recalculated.

Procedure

1. Construct the multipurpose fourth-order filter circuit shown in Figure 34-2 (jumper J_1 connected).
2. Sketch the gain-versus-frequency response curves for the bandpass (BP_b) and bandstop (N_b) filter outputs on semilog graph paper.
3. From the response curves sketched in step 2, determine the center frequencies for the bandpass and bandstop filters and their respective bandwidths and Q factors.
4. Compare the results measured in step 3 to the values measured in Section A.
5. Vary the clock frequency above and below 100 kHz, and observe the frequency response characteristics.
6. Describe what effect varying the clock frequency has on the response curves for the bandpass and bandstop filters.

SECTION C   Summary

Write a brief summary of the concepts presented in this experiment on digital filters. Include the following items:

1. The basic operation of a switched capacitor digital filter.
2. The relationship between the external clock frequency, the center frequency of the bandpass and bandstop filters, and the break frequency for the lowpass filter.
3. The relationship between the external resistance values and the gains of the various filter outputs.
4. The advantages of using monolithic integrated circuit filters over conventional discrete component filters.

EXPERIMENT 34 ANSWER SHEET

NAME: _____ CLASS: _____ DATE: _____

SECTION A

2. F_o = _____   3. Q = _____
4. B = _____   5. H_{olp} = _____
6. H_{obp} = _____   7. H_{on} = _____
9. Bandpass filter, H_{obp} = _____ F_c = _____ B = _____ Q = _____
   Bandstop filter, H_{on} = _____ F_c = _____ B = _____ Q = _____
   Lowpass filter, H_{olp} = _____ F_U = _____

10. _____

_____

_____

13. _____

_____

_____

15. _____

_____

_____

SECTION B

3. Bandpass filter, F_c = _____ B = _____ Q = _____
   Bandstop filter, F_c = _____ B = _____ Q = _____

4. _____

_____

_____

6. _____

_____

_____

SAMPLE-AND-HOLD CIRCUIT

REFERENCE TEXT: Electronic Communications Systems: Fundamentals through Advanced

1. Chapter 13, 8PSK Transmitter.
2. Chapter 16, Sample-and-Hold Circuit.
3. Chapter 16, Sampling Rate.

OBJECTIVES

1. To observe the operation of a sample-and-hold circuit.
2. To observe the operation of a linear integrated-circuit sample-and-hold circuit.
3. To observe pulse amplitude modulation.

INTRODUCTION

The function of a sample-and-hold circuit is to sample periodically a continually changing analog input signal and convert the samples to voltage levels. In essence, a sample-and-hold circuit converts analog signals to a pulse amplitude-modulated wave. With pulse amplitude modulation (PAM), an analog signal is sampled and converted to a series of constant-amplitude PAM levels. In this experiment the operation of the LF398A monolithic sample-and-hold circuit is examined. The LF398A is a monolithic sample-and-hold integrated circuit which utilizes BI-FET technology to obtain ultra-high dc accuracy with fast acquisition of signal and a low droop rate. The LF398A features a TTL, PMOS, and CMOS compatible logic input; less than 10 μs acquisition time; low input offset; low output noise in hold mode; and wide bandwidth. The functional block diagram for the LF398A is shown in Figure 35-1a, and the DIP pin layout is shown in Figure 35-1b. A bipolar input stage is used to achieve low offset voltage and a wide bandwidth, which allows the LF398A to be used inside the feedback loop of a 1 MHz op-amp circuit without stability problems. P-channel JFETs are combined with bipolar devices in the output amplifier to give droop rates as low as 5 mV/min with a 1 μF external holding capacitor.

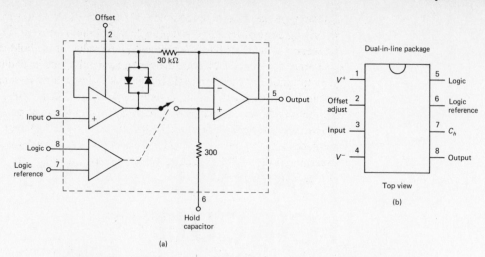

FIGURE 35-1   LF398A Sample-and-hold circuit. (a) Functional block diagram. (b) DIP pin out.

MATERIALS REQUIRED

Equipment:

1 — protoboard

1 — dual dc power supply (+15 V dc and − 15 V dc)

1 — variable dc power supply (− 10 V dc to + 10 V dc)

1 — medium-frequency function generator (100 kHz)

1 — low-frequency audio signal generator (20 kHz)

1 — standard oscilloscope (10 MHz)

1 — assortment of test leads and hookup wire

Parts list:

1 — LF398A monolithic sample-and-hold circuit

1 — CD4069 hex inverter

1 — 1k-ohm variable resistor

1 — 10k-ohm variable resistor

1 — 0.001 μF-capacitor

2 — 0.01 μF-capacitors

1 — 0.05 μF-capacitor

1 — 0.1-μF-capacitor

SECTION A   The Sample-Pulse Generator

In this section the operation of a sample-pulse generator is examined. The Nyquist sampling theorem establishes the minimum sampling rate (F_s) that can be used for a given sample-and-hold circuit. It states that for a sample to be reproduced accurately at the receiver, each cycle of the analog input signal (F_a) must be sampled at least twice. Consequently, the minimum sampling rate is equal to twice the highest input frequency. If F_s is less than twice F_a, foldover distortion (sometimes called aliasing) occurs. In addition, monolithic sample-and-hold circuits, such as the LF398A, impose limitations on the rise time of the input logic signal (sample pulse). For proper operation, sample pulses into the LF398A must have a minimum d_V/d_t of 1.0 V/μs. In essence, the LF398A has two operating modes: the sample mode and the hold mode. The sample mode is initiated on the rising edge of the input sample pulse; and the hold command is internally generated between 4 and 20 μs later, depending on the value of the hold capacitor. Very often the rise times associated with square wave sample pulses are inadequate for proper operation. One technique of rectifying this problem is to insert a wave-shaping circuit between the square wave generator and the sample clock input to the LF398A. The schematic diagram for the sample pulse wave-shaping network used in this section is shown in Figure 35-2. A square-wave sample pulse is generated by the function generator. The RC differentiator produces a narrow nonrectangular pulse which is squared off and cleaned up by the two inverters in the CD4069. The RC time constant, and thus the duty cycle of the sample pulse, is adjusted with R_x.

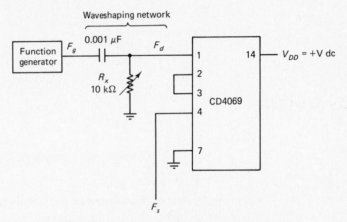

FIGURE 35-2   Sample pulse wave-shaping network.

Procedure

1. Construct the wave-shaping circuit shown in Figure 35-2 (set R_x to midrange).
2. Set the amplitude of the function-generator output voltage to a 10 Vp-p square wave with a frequency $F_g = 10$ kHz.
3. Sketch the waveforms observed at the following points: the function generator output (F_g), the RC differentiator output (F_d), and the second inverter output (F_s).
4. Vary R_x throughout its entire range while observing F_s and F_d.
5. Describe what effect varying R_x has on the shape of F_s and F_d.
6. Vary the amplitude of the function generator output voltage.
7. Describe what effect varying the function generator output voltage has on F_s and F_d.
8. Readjust the amplitude of the function generator output voltage to 10 Vp-p and reset R_x to midrange.

SECTION B   The Linear Integrated Circuit Sample-and-Hold Circuit

In this section the operation of the LF398A monolithic sample-and-hold circuit is examined. The schematic diagram for the sample-and-hold circuit used in this section is shown in Figure 35-3. The only external component required for the LF398A is a hold capacitor. Hold step, acquisition time, and droop rate are the major trade-offs in the selection of the hold capacitor. Hold step is the voltage step at the output of the sample-and-hold circuit when switching from the sample to the hold mode of operation with a steady (dc) analog

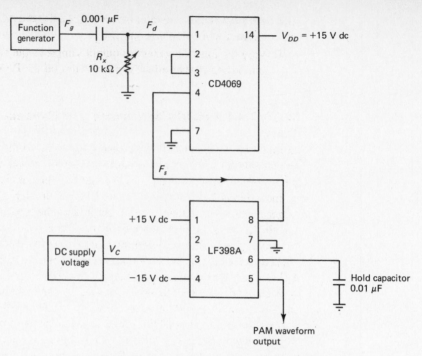

FIGURE 35-3   LF398A Sample-and-hold circuit.

input voltage. Acquisition time is the time required to acquire a new analog input voltage with an output step of 10 V. Droop rate is the change in output voltage per unit of time that occurs while the circuit is in the hold mode. Several curves are included in the specification sheet to aid in the selection of a reasonable value of hold capacitance.

Procedure

1. Construct the sample-and-hold circuit shown in Figure 35-3.
2. Set the dc voltage source (V_C) to 0 V dc, and set the amplitude of the function-generator output voltage to a 10 Vp-p square wave with a frequency $F_g = 10$ kHz.
3. With an oscilloscope, observe the analog input voltage to the sample-and-hold circuit and the PAM output voltage.
4. Adjust R_x for a PAM output waveform with minimum distortion.
5. Measure the duty cycle of the sample pulse (F_S).
6. Vary the dc supply voltage (V_C) from -10 V dc to $+10$ V dc while observing the PAM output voltage.
7. What are the maximum positive and maximum negative input voltages that the sample-and-hold circuit will accurately convert to a PAM output signal?
8. Replace the dc supply voltage with an audio signal generator, and set the amplitude of the signal generator output voltage to a 4 V-p-p sine wave with a frequency $F_a = 1$ kHz.
9. Fine tune the audio signal generator output amplitude and frequency to achieve a stable PAM output waveform.
10. Vary the signal generator output frequency between 100 Hz and 4 kHz and describe what effect varying it has on the PAM output waveform.
11. Determine the maximum input frequency that the sample-and-hold circuit will remain locked to by increasing the signal generator output frequency until the PAM waveform no longer follows the input signal.

12. Vary the signal generator output amplitude, and describe what effect varying it has on the PAM output waveform.

13. Vary the function generator output sample frequency between 5 kHz, and 20 kHz, and describe what effect varying it has on the PAM output waveform.

SECTION C   Hold Capacitance and Dynamic Sampling

In this section the effects of increasing the value of the hold capacitor for the LF398A sample-and-hold circuit are examined. When the sample-and-hold circuit is in the sample mode, the hold capacitor charges. The rate at which it charges is determined by the value of the external hold capacitor, which charges through a 300-ohm internal resistor. The larger the hold capacitance, the longer it takes the capacitor to charge and thus the lower the amplitude of the PAM output voltage. The acquisition time (i.e., the time the capacitor is charging) is inversely proportional to the hold capacitance. Also, increasing the hold capacitance increases the switching delay time between the sample and hold modes. The RC charging time constant can also be increased by placing an external resistor in series with the hold capacitor. Sample errors due to a moving input signal produce distortion in the PAM output signal and can cause the sample-and-hold circuit to lose lock. This is partially due to the internal time delay between the rising edge of the input logic signal and the internally generated hold command. Large capacitance values produce a longer internal time delay and thus a more distorted output PAM signal and a higher probability of losing lock. The schematic diagram for the sample-and-hold circuit used in this section is shown in Figure 35-4.

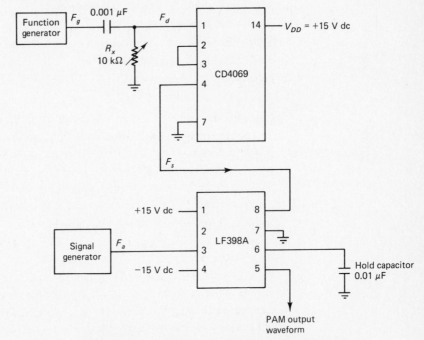

FIGURE 35-4   LF398A Sample-and-hold, dynamic sampling.

Procedure

1. Construct the sample-and-hold circuit shown in Figure 35-4, and adjust R_x to its minimum resistance value (0 ohms).

2. Set the amplitude of the function generator output voltage to a 10 Vp-p square wave with a frequency $F_g = 10$ kHz.

3. Set the amplitude of the signal generator output voltage to a 4 Vp-p sine wave with a frequency $F_a = 1$ kHz.

4. Adjust R_x for a PAM output signal with minimum distortion (it may be necessary to fine tune the signal-generator output amplitude and frequency to achieve a stable PAM output waveform).

5. Sketch the PAM output waveform.

6. Replace the hold capacitor with a 0.05 μF capacitor, and repeat steps 4 and 5.

7. Replace the hold capacitor with a 0.1 μF capacitor, and repeat steps 4 and 5.

8. Describe what effect increasing the hold capacitance has on the PAM output waveform.

9. Replace the hold capacitor with a 0.01 μF capacitor.

10. Slowly increase the resistance of R_x while observing the PAM output waveform.

11. Describe what effect increasing the resistance of R_x has on the PAM output waveform.

SECTION D   Summary

Write a brief summary of the concepts presented in this experiment on the sample-and-hold circuit. Include the following items:

1. The Nyquist sample rate.
2. Foldover distortion.
3. Acquisition time, hold time, and droop rate.
4. Sample and hold modes.
5. Sample-and-hold switching delay.
6. Sample-and-hold frequency lock.
7. Pulse amplitude modulation.

EXPERIMENT 35 ANSWER SHEET

NAME: _____ CLASS: _____ DATE: _____

SECTION A

3.

Vertical sensitivity _____ V/cm
Time base _____ sec/cm

Vertical sensitivity _____ V/cm
Time base _____ sec/cm

Vertical sensitivity _____ V/cm
Time base _____ sec/cm

5. _____
_____
_____

7. _____
_____
_____

SECTION B

5. Duty cycle = _____

7. Maximum positive voltage = _____
Maximum negative voltage = _____

10. _____

_____

_____

11. Maximum input frequency = _____

12. _____

_____

_____

13. _____

_____

_____

SECTION C

5.

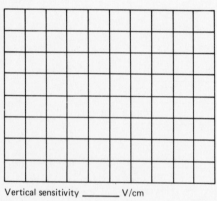

Vertical sensitivity _____ V/cm

Time base _____ sec/cm

6.

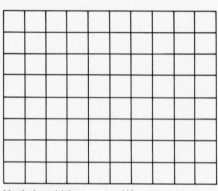

Vertical sensitivity _____ V/cm

Time base _____ sec/cm

7.

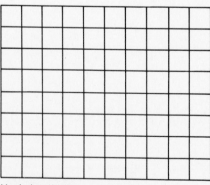

Vertical sensitivity _____ V/cm

Time base _____ sec/cm

8. _____

_____

_____

11. _____

_____

_____

TIME-DIVISION MULTIPLEXING

REFERENCE TEXT: Electronic Communications Systems: Fundamentals through Advanced

1. Chapter 17, Time-Division Multiplexing.

OBJECTIVES

1. To observe the operation of a time-division multiplexer.
2. To observe the operation of a time-division demultiplexer.
3. To observe the operation of a time-division multiplexing system.

INTRODUCTION

Multiplexing is the transmission of information in either analog or digital form from more than one source to more than one destination on the same transmission medium, such as a pair of wires. Transmissions occur on the same facility but not necessarily at the same time. With time-division multiplexing (TDM), transmissions from multiple sources occur on the same facility but not at the same time. Transmissions from various sources are interleaved in the time domain. The transmissions from each source occupy separate time slots (epochs) within the total TDM frame. Multiplex simply means many to one. In the demultiplexer, a TDM transmission is separated into parts, and the information from each part is connected to its appropriate receiver. Demultiplex simply means one to many. With TDM, if the original source information is analog, it must be converted to digital form prior to transmission and then converted back to analog form at the receiver. If the original source information is digital, it must be converted to an appropriate transmission (bit) rate prior to transmission. In this experiment two binary information sources are time-division multiplexed onto a common wire, transmitted, and then demultiplexed in the receiver. Either source can be connected to either receiver with the appropriate logic circuitry. The block diagram for the time-division multiplex/demultiplex system used in this experiment is shown in Figure 36-1.

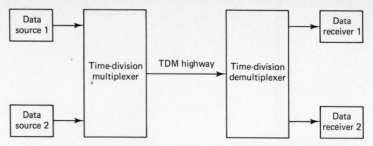

FIGURE 36-1   Time-division multiplexer/demultiplexer system.

MATERIALS REQUIRED

Equipment:

1 — protoboard
1 — dc power supply (+12 V dc)
1 — standard oscilloscope (10 MHz)
1 — assortment of test leads and hookup wire

Parts list:

1 — CD4081 quad two-input AND gate
1 — CD4071 quad two-input OR gate
1 — CD4027B dual JK flip-flop
1 — CD4069B hex inverter
1 — Dual SPDT switch
2 — 1k-ohm resistors

SECTION A   Channel 1 and Channel 2 Data Sources

In this section two data sources are constructed. One data source is for channel 1 and one data source is for channel 2. In subsequent sections the two data sources are time-division multiplexed onto TDM data channels 1 and 2 and transmitted over a metallic wire to the demultiplexer. In the demultiplexer the two data channels are separated and fed to their respective receivers. For simplicity, a function generator is used to generate a 1000-bps alternating 1/0 data pattern (i.e., a 500-Hz square wave) for the channel 1 data source. The channel 1 data source is simply divided by two to generate a second 1000-bps data source except with a repeating 1100 data pattern (i.e., a 250-Hz square wave). The second data channel is further divided by 2 to produce a 125-Hz square wave which is used to simulate the sample pulse in both the multiplexer and the demultiplexer. The 125-Hz square wave is used to alternately select 4 bits from each data source. The schematic diagram for the data source generator used in this section is shown in Figure 36-2a and the corresponding timing diagram is shown in Figure 36-2b.

Procedure

1. Construct the data source generator circuit shown in Figure 36-2a.
2. Set the amplitude of the function-generator output voltage to a 10 Vp-p square wave with a maximum positive voltage $V_{max} = +10$ V and a maximum negative voltage $V_{min} = 0$ V.

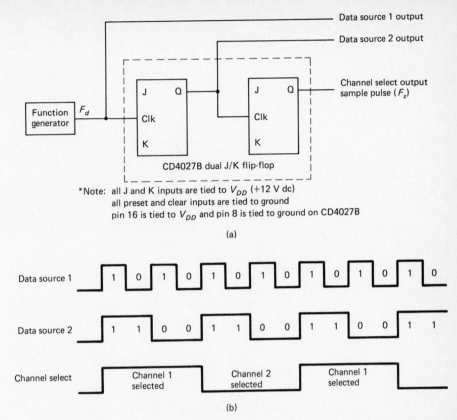

FIGURE 36-2   Data source generator. (a) Schematic diagram. (b) Timing diagram.

3. Set the function generator output frequency (F_d) to 500 Hz.

4. Sketch the waveforms at the data source 1 and data source 2 outputs.

5. Describe the waveforms sketched in step 4.

6. Sketch the waveform observed at the channel-select output.

7. Describe the relationship between the waveform sketched in step 6 and the waveforms sketched in step 4.

8. Do not disassemble this circuit, as it is needed in subsequent sections.

SECTION B   Time-Division Multiplexer

In this section the operation of a two-channel time-division multiplexer is examined. The schematic diagram for the two-channel multiplexer circuit used in this section is shown in Figure 36-3. Data sources 1 and 2 are simulated with the circuit constructed in Section A. With time-division multiplexing, either data source 1 or data source 2 can be externally selected and transferred to the output TDM highway. Source 1 and source 2 are selected with the MUX channel-select switch S_1. Whichever channel is selected appears at the multiplexer output until the position of S_1 is changed. If S_1 is replaced with a 125-Hz square wave sample pulse (F_s), data bits from sources 1 and 2 are alternately switched to the TDM output in groups of 4 bits. Four bits from source 1 (1010) occupy the time slots (epochs) within the total TDM frame for channel 1, and 4 bits (1100) from data source 2 occupy the epochs for channel 2. Thus, a repetitive 10101100 pattern appears on the TDM output highway.

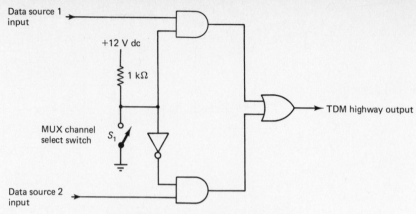

FIGURE 36-3   Time-division multiplexer.

Procedure

1. Construct the 2-channel time-division multiplexer circuit shown in Figure 36-3.
2. Place channel selector switch S_1 in the open position, and observe the waveform at the TDM output.
3. Sketch the waveform observed in step 2.
4. Place channel selector switch S_1 in the closed position, and observe the waveform at the TDM output.
5. Sketch the waveform observed in step 4.
6. Describe the waveforms sketched in steps 3 and 5.
7. Remove channel selector switch S_1, and connect the channel select output from the data source circuit to the MUX channel select input on the multiplexer.
8. Sketch the waveform observed at the TDM output.
9. Describe the waveform sketched in step 8.
10. Do not disassemble this circuit, as it is needed in subsequent sections.

SECTION C   The Time-Division Demultiplexer

In this section the operation of a two-channel time-division demultiplexer is examined. The schematic diagram for the two-channel demultiplexer circuit used in this section is shown in Figure 36-4. Data sources 1 and 2 are time-division multiplexed onto the TDM highway and transmitted to the demultiplexer. In the demultiplexer, DEMUX-channel selector switch S_2 is used to ''steer'' data from the TDM highway to either data receiver 1 or data receiver 2. S_1 and S_2 can be used to connect data source 1 or data source 2 to either data receiver 1 or data receiver 2. If S_2 is replaced with a 125-Hz square-wave sample pulse (F_s), data bits from the TDM highway are alternately switched to receiver 1 and receiver 2 in groups of 4 bits. If the sample pulse in the demultiplexer is synchronized with the sample pulse in the multiplexer, data is transferred from source 1 to data receiver 1 and from source 2 to data receiver 2 in groups of 4 bits.

Procedure

1. Construct the two-channel multiplexer/demultiplexer circuit shown in Figure 36-4.
2. Place channel selector switches S_1 and S_2 in their open positions.
3. Observe the waveforms at the following locations: data source 1 input, data source 2 input, TDM output highway, data receiver 1 output, and data receiver 2 output.

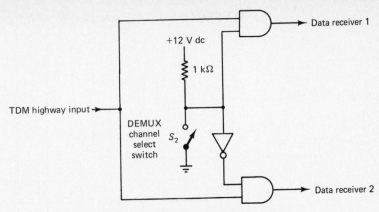

FIGURE 36-4   Time-division demultiplexer.

4. Describe the relationship between the waveforms observed in step 3.
5. Switch channel selector switch S_1 to its closed position, and repeat steps 3 and 4.
6. Place channel selector switches S_1 in the open position and channel-selector switch S_2 in the closed position, and repeat steps 3 and 4.
7. Change S_2 to its open position, and repeat steps 3 and 4.
8. Remove channel selector switches S_1 and S_2, and connect the channel-select output from the data source circuit to both the MUX and DEMUX channel select inputs on the multiplexer and demultiplexer, respectively; and repeat steps 3 and 4.

SECTION D   Summary

Write a brief summary of the concepts presented in this experiment on time-division multiplexing and demultiplexing. Include the following items:

1. The concept of multiplexing.
2. The concept of demultiplexing.
3. The concept of time-division multiplexing.
4. The terms *frame* and *frame time*.
5. The terms *epoch* and *sample pulse*.

EXPERIMENT 36 ANSWER SHEET

NAME: _____ CLASS: _____ DATE: _____

SECTION A

4.

Vertical sensitivity _____ V/cm   Vertical sensitivity _____ V/cm

Time base _____ sec/cm   Time base _____ sec/cm

5. _____

_____

_____

6.

Vertical sensitivity _____ V/cm

Time base _____ sec/cm

7. _____

_____

_____

SECTION B

3.

Vertical sensitivity _____ V/cm

Time base _____ sec/cm

5.

Vertical sensitivity _____ V/cm

Time base _____ sec/cm

6. _____

_____

_____

8.

Vertical sensitivity _____ V/cm

Time base _____ sec/cm

9. _____

_____

_____

SECTION C

4. _____

_____

_____

5. _____

_____

_____

6. _____

_____

_____

7. _____

_____

_____

8. _____

_____

_____

Experiment 37

PCM-TDM SYSTEM DESIGN USING COMBO CHIPS

REFERENCE TEXT: Electronic Communications Systems: Fundamentals through Advanced

1. Chapter 17, Codecs
2. Chapter 17, 2913/14 Combo Chip.

OBJECTIVES

1. To observe the operation of a PCM-TDM multiplexer/demultiplexer system.
2. To observe the operation of a large-scale-integration PCM-TDM multiplexer/demultiplexer system.
3. To observe the operation of a large-scale-integration combo chip.
4. To observe the operation of a codec.

INTRODUCTION

A codec is a large-scale-integration (LSI) chip designed for use in the telecommunications industry for private branch exchanges (PBXs), central office switches, digital handsets, voice store-and-forward systems, and digital echo suppressors. Essentially, a codec is applicable for any purpose that requires the digitizing of analog signals, such as in a PCM-TDM carrier system. *Codec* is a generic term that refers to the coding and decoding functions performed by a device that converts analog signals to digital codes and digital codes to analog signals. Recently developed codecs are called combo chips because they combine codec and filter functions in the same LSI package. In this experiment a simplex single- and two-channel PCM system will be designed, constructed, and tested to meet Bell System T1 carrier specifications. The block diagrams and designations used in this experiment apply to the 2913/14 combo chip, although any comparable combo chip can be used. The 2913/14 is a combo chip that can provide the analog-to-digital and digital-to-analog conversions and the transmit and receive filtering necessary to interface a full-duplex voice telephone circuit to the PCM highway of TDM carrier system. The following major functions are provided by the 2913/14 combo chip:

1. Bandpass filtering of the analog signals prior to encoding and after decoding (i.e., anti-alias filtering).
2. Encoding and decoding of voice and call-progress signals.
3. Encoding and decoding of signaling and supervision information.
4. Digital companding.

The functional block diagram for the 2914 combo chip is shown in Figure 37-1.

MATERIALS REQUIRED

Equipment:

1 — protoboard
1 — dc power supply (+ V dc)
1 — audio-signal generator (10 KHz)
1 — standard oscilloscope (10 MHz)
1 — assortment of test leads and hookup wire

Parts list:

1 — 2913/14 combo chip or equivalent
1 — assortment of resistors, capacitors, and integrated circuits depending on the individual circuit implementation.

SECTION A   The Single-Channel PCM System — the Fixed-Data-Rate Mode

In this section the operation of a single-channel PCM system operating in the fixed-data-rate mode is examined. In the fixed-data-rate mode, the master transmit and receive clocks (CLKX and CLKR, respectively) perform the following functions:

1. Provide the master clock for the on-board switched capacitor anti-alias filters.
2. Provide the synchronizing clock for the analog-to-digital and digital-to-analog converters.
3. Determine the input and output data rates between the combo chip and the PCM highway. In the fixed-data-rate mode, the transmit and receive data rates must be either 1.536, 1.544, or 2.048 Mbps, the same as the master clock rate.

The block diagram for the single-channel PCM system used in this section is shown in Figure 37-2a, and the timing diagram is shown in Figure 37-2b. The master-clock frequency must be either 1.536, 1.544, or 2.048 MHz. In the fixed data-rate mode, data are inputted and outputted in short bursts. With only a single channel, the PCM highway is active only 1/24 of the total frame time. Transmit and receive frame synchronizing pulses (FSX and FSR) are 8-kHz clock inputs which set the transmit and receive sampling rates and distinguish between signaling and nonsignaling frames. Supervisory signaling will not be used in this experiment. Therefore, the SIGX input is permanently tied to ground, and the SIGR output is left open. An 8-kHz sampling rate establishes the maximum audio input frequency at 4 kHz. $\overline{TSX}$ is a time-slot strobe buffer enable output which is used to gate the PCM word onto the PCM highway when an external buffer is used to drive the line. $\overline{TSX}$ is also used as an external gating pulse when additional channels are time-division multiplexed. Data are transmitted to the PCM highway from DX on the first eight positive transitions of CLKX following the rising edge of FSX. On

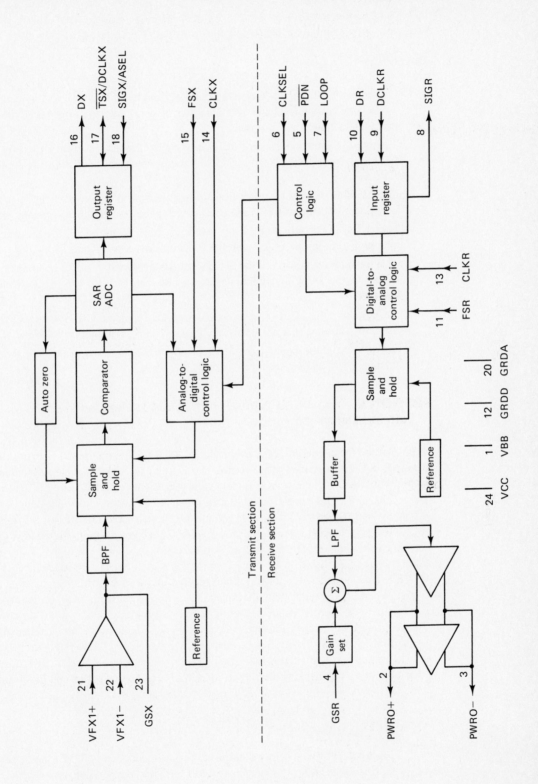

the receive channel, data are received from the PCM highway from DR on the first eight falling edges of CLKR after the occurrence of FSR. Therefore, the occurrence of FSX and FSR must be synchronized between codecs in a multichannel system to ensure that only one codec is transmitting to, or receiving from, the PCM highway at any given time.

Procedure

1. Construct the single-channel fixed-data-rate mode PCM system shown in Figure 37-2a. Use components and additional ICs of your choosing to generate the appropriate clocks and timing signals.

2. Test the transmitter portion of the PCM system constructed in step 1 to ensure that for each sample taken 8 bits of PCM code are transmitted during the appropriate time slot.

3. Troubleshoot the PCM transmitter as required, and make any design changes, adaptions, or additions that are necessary for the transmitter to operate properly.

4. Sketch the PCM output signal observed on the PCM highway, the transmit frame synchronizing pulse (FSX), and the transmit time-slot strobe (TSX).

5. Test the receiver portion of the PCM system constructed in step 1 by connecting the output of the PCM transmitter tested in step 2 directly to the input of the PCM receiver.

6. Troubleshoot the PCM receiver as required, and make any design changes, adaptions, or additions that are necessary for the receiver to operate properly.

7. On semilog graph paper, plot the audio-frequency response curve for the single-channel PCM system tested in steps 1 through 6.

SECTION B   The Two-Channel PCM-TDM system — the Variable Data Rate Mode

In this section the operation of a two-channel PCM-TDM system operating in the variable-data-rate mode is examined. The variable-data-rate mode allows for a flexible data input and output clock frequency. It provides the ability to vary the frequency of the transmit and receive clock rates. In the variable-data-rate mode, a master-clock frequency of either 1.536, 1.544, or 2.048 MHz is still required for proper operation of the on-board band-pass filters and the analog-to-digital and digital-to-analog converters. The block diagram for the two-channel PCM system used in this section is shown in Figure 37-3a, and the timing diagram is shown in Figure 37-3b. In the variable-data-rate mode, DCLKR and DCLKX become the data clocks for the receive and transmit PCM highways, respectively. When FSX is high, data are transmitted onto the PCM highway on the next eight consecutive positive transitions of DCLKX. Similarly, while FSR is high, data from the PCM highway are clocked into the codec on the next eight consecutive negative transitions of DCLKR. The transmit and receive enable signals (FSX and FSR) for each codec are active for one half of the total frame time. Consequently, 8-kHz 50% duty cycle transmit and receive data enable signals are fed directly to one codec and fed to the other codec inverted, thereby enabling only one codec at a time.

Procedure

1. Construct the two-channel variable-data-rate mode PCM-TDM system shown in Figure 37-3a. Use components and additional ICs of your choosing to generate the appropriate clocks and timing signals.

2. Test the two codecs and the TDM multiplexer of the PCM-TDM system constructed in step 1 to ensure that each channel is transmitting 8 bits of PCM code for each sample taken within the appropriate time slot.

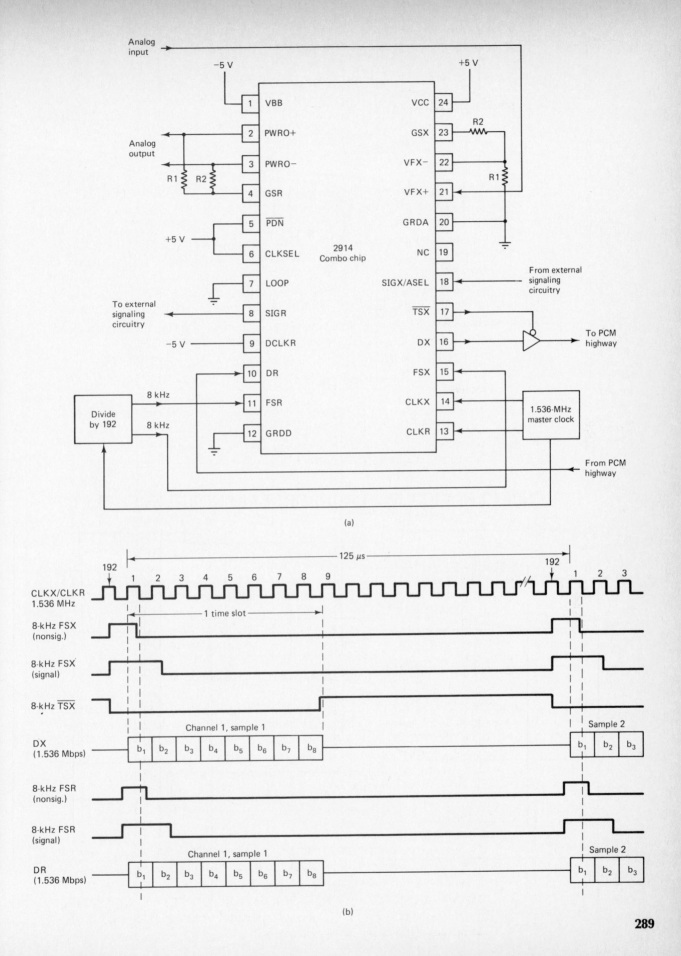

(a)

(b)

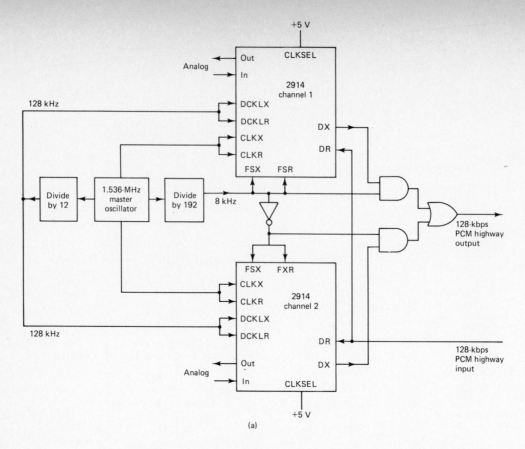

(a)

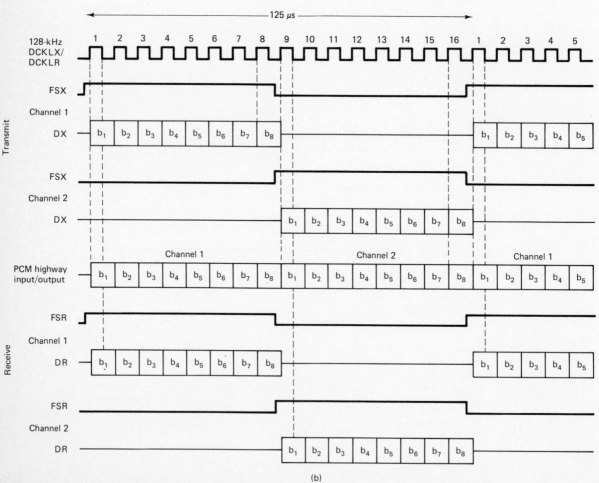

(b)

3. Troubleshoot the codecs and the TDM multiplexer as required, and make any design changes, adaptions, or additions that are necessary for their proper operation.

4. Sketch the output signal observed on the PCM-TDM highway, the transmit frame synchronizing pulses, and the transmit data rate clocks.

5. Test the demultiplexer and receiver portion of the PCM-TDM system constructed in step 1 by connecting the output of the time-division multiplexer tested in step 2 directly to the PCM receiver.

6. Troubleshoot the PCM-TDM receiver as required, and make any design changes, adaptions, or additions that are necessary for the receiver to operate properly.

7. On semilog graph paper, plot the audio frequency response curves for both channels of the two-channel PCM/TDM system tested in steps 2 through 6.

SECTION C   Summary

Write a complete laboratory report for the single- and two-channel PCM-TDM systems constructed and tested in Sections A and B. Include the following items:

1. Complete schematic diagrams of both systems.
2. An introduction including a statement of the problem.
3. All necessary calculations.
4. A list of problems encountered and how they were rectified.
5. A conclusion.

Experiment 38

PCM-TDM LINE ENCODING

REFERENCE TEXT: Electronic Communications Systems: Fundamentals through Advanced

1. Chapter 17, Line Encoding.

OBJECTIVES

1. To observe the operation of several PCM-TDM line encoders.
2. To observe the transmission characteristics of the following line encoding schemes: UPNRZ, BPNRZ, UPRZ, and BPRZ.

INTRODUCTION

Line encoding involves converting standard logic levels (i.e., TTL, CMOS, etc.) to a form more suitable for telephone-line transmission. Essentially, there are four primary factors that must be considered when selecting a line encoding format.

1. Timing (clock) recovery. With PCM-TDM carrier systems, clocking information is generally recovered directly from the received data. Therefore, there must be a sufficient number of transitions in the data signal to allow for continuous and accurate clock recovery.

2. Transmission bandwidth. The minimum bandwidth required to propagate a digital line encoded signal through a transmission medium is determined by the highest fundamental frequency present in the waveform. The highest fundamental frequency is determined from the worst-case (fastest-transition) binary bit sequence. The highest fundamental frequency is the minimum bandwidth and is called the minimum or ideal Nyquist frequency.

3. Ease of detection and decoding. The dc component associated with a digital line encoding scheme and the transmit voltage levels are dependent on the particular line-encoding scheme used. Unipolar transmission transmits 0 V for a logic O and either a

positive or a negative voltage for a logic 1. Bipolar transmission transmits both positive and negative voltages for logic 1s and 0 V or a negative voltage for logic 0s. The average dc component for a digital transmission is dependent on whether unipolar or bipolar encoding is used.

4. Error detection. With the rapidly increasing transmission bit rates, error detection has become an important consideration when selecting a line encoding scheme.

In this experiment four types of line-encoding schemes are examined: unipolar nonreturn-to-zero (UPNRZ), bipolar nonreturn-to-zero (BPNRZ), unipolar return-to-zero (UPRZ), and bipolar return-to-zero (BPRZ). Figure 38-1 shows the timing diagrams for the four line encoding schemes used in this experiment. To simplify circuit implementation, only an alternating 1/0 binary data input is used in this experiment. To implement encoders that will function for any binary input sequence, additional circuitry and an external clock are necessary.

MATERIALS REQUIRED

Equipment:

1 — protoboard
1 — dual dc power supply (+15 V dc and −15 V dc)
1 — variable dc power supply (−5 V dc to +5 V dc)

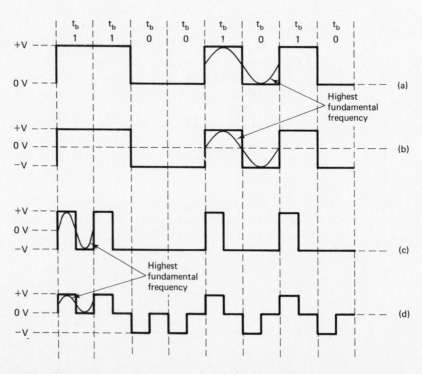

FIGURE 38-1   Line-encoding formats: (a) UPNRZ; (b) BPNRZ; (c) UPRZ; (d) BPRZ.

1 — low-frequency function generator (4000 Hz)

1 — standard oscilloscope (10 MHz)

1 — assortment of test leads and hookup wire

Parts list:

1 — LS123 dual retriggerable monostable multivibrator

1 — op amp (741 or equivalent)

1 — CD4069 hex inverter

1 — 4.7k-ohm resistor

4 — 10k-ohm resistors

2 — 10k-ohm variable resistors

2 — 0.1-uF capacitors

SECTION A   The Unipolar Nonreturn-to-Zero to Bipolar Nonreturn-to-Zero Line Encoder

In this section the transmission characteristics of unipolar nonreturn-to-zero (UPNRZ) and bipolar nonreturn-to-zero (BPNRZ) line encoders are examined. With unipolar transmission, a single nonzero voltage level represents a logic 1 (i.e., either $+V$ or $-V$, depending on whether positive or negative logic is used), and zero volts represents a logic 0. For example, with standard TTL, a logic $1 = +5$ V and a logic $0 =$ ground. With bipolar transmission, a logic 1 can be represented by both positive and negative voltages (i.e. logic $1 = +V$ or $-V$), and zero volts represents a logic 0; or logic 1s are represented with a positive voltage $(+V)$, and logic 0s represented with a negative voltage $(-V)$. With nonreturn-to-zero transmission, the level of the binary pulse is maintained for the entire bit time (i.e., a 100% duty cycle), and with return-to-zero transmission the active time of the binary pulse is less than 100% of the bit time (generally, a 50% duty cycle is used). UPNRZ transmission is the standard binary transmission scheme used internally with computers and conventional digital equipment. Sample timing diagrams for UPNRZ and BPNRZ transmission are shown in Figures 38-1a and 38-1b, respectively. The schematic diagram for the UPNRZ to BPNRZ line encoder circuit used in this section is shown in Figure 38-2.

Procedure.

1. Construct the UPNRZ to BPNRZ line encoder circuit shown in Figure 38-2.

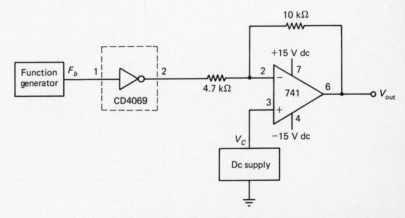

FIGURE 38-2   UPNRZ encoder circuit.

2. Set the dc bias supply (V_C) to -1.5 V dc.

3. Use the function generator to simulate a 2000 bps UPNRZ input signal (F_b) with an alternating 1/0 sequence by setting the amplitude of the function-generator output voltage to a 5 Vp-p 1000-Hz square wave ($V_{max} = +5$ V and $V_{min} = O$ V).

4. Observe the waveform at V_{out}.

5. Adjust the dc bias supply (V_c) until a symmetrical BPNRZ output waveform is observed at V_{out}.

6. Sketch the input UPNRZ and output BPNRZ waveforms.

7. Measure the input voltage level for a logic 1. A logic 0.

8. Measure the output voltage level for a logic 1. A logic 0.

9. Measure the average dc component of the input waveform. The output waveform.

10. Determine the ideal Nyquist Frequency for the input waveform. The output waveform.

SECTION B The Unipolar Nonreturn-to-Zero-to-Unipolar Return-to-Zero Line Encoder

In this section the transmission characteristics of unipolar nonreturn-to-zero (UPNRZ) and unipolar return-to-zero (UPRZ) line encoders are examined. One half of an LS123 retriggerable monostable multivibrator is used in the encoder to generate a variable-duty-cycle unipolar return-to-zero waveform. The LS123 is a dual retriggerable monostable multivibrator with output pulse-width control. In this section the pulse width of the return-to-zero output waveform is programmed by selection of external resistance (R_{ext}) and capacitance (C_{ext}) values. The schematic diagram for the UPNRZ to UPRZ line encoder circuit used in this section is shown in Figure 38-3, and sample timing diagrams for UPNRZ and UPRZ transmission are shown in Figure 38-1a and c, respectively.

Procedure

1. Construct the UPNRZ to UPRZ line encoder circuit shown in Figure 38-3.

2. Use the function generator to simulate an 8000 bps UPNRZ input signal (F_b) with an alternating 1/0 sequence by setting the amplitude of the function-generator output voltage to a 5 Vp-p 4000-Hz square wave ($V_{max} = 5$ V and $V_{min} = 0$ V).

3. Observe the waveform at V_{out}.

4. Adjust R_{ext} until a 50%-duty-cycle UPRZ waveform is observed at V_{out}. (Note that

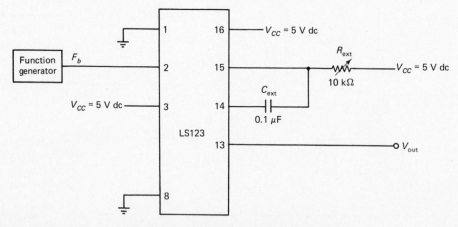

FIGURE 38-3   UPNRZ to UPRZ encoder circuit.

with an alternating 1/0 input sequence, the output waveform will be a 25% duty cycle retangular waveform.)

5. Sketch the input UPNRZ and output UPRZ waveforms.
6. Measure the input voltage level for a logic 1. A logic 0.
7. Measure the output voltage level for a logic 1. A logic 0.
8. Measure the average dc component of the input waveform. The output waveform.
9. Vary R_{ext}, and describe what effect varying it has on the shape and dc component of the output waveform.
10. Determine the ideal Nyquist frequency for the input waveform. The output waveform.

SECTION C   The Unipolar Nonreturn-to-Zero to Bipolar Return-to-Zero Line Encoder

In this section the transmission characteristics of unipolar nonreturn-to-zero (UPNRZ) and bipolar return-to-zero (BPRZ) are examined. Both halves of an LS123 dual retriggerable monostable multivibrator are used in the encoder to generate a variable duty cycle bipolar return-to-zero waveform. With BPRZ transmission a 50% duty cycle positive voltage is transmitted for a logic 1 signal, and a 50% duty cycle negative voltage is transmitted for a logic 0 signal. The A-section of the LS123 is used to generate a 50% duty cycle waveform for the logic 1 UPNRZ input signals, and the B-section of the LS123 is used to generate a 50% duty cycle waveform for the logic 0 UPNRZ input signals. The return-to-zero waveform for the logic 0 input signals are inverted and then added to the logic 1 return-to-zero waveform in the op amp to produce a bipolar return-to-zero output waveform. The schematic diagram for the UPNRZ-to-BPRZ encoder circuit used in this section is shown in Figure 38-4, and sample timing diagrams for UPNRZ and BPRZ transmission are shown in Figure 1a and d, respectively.

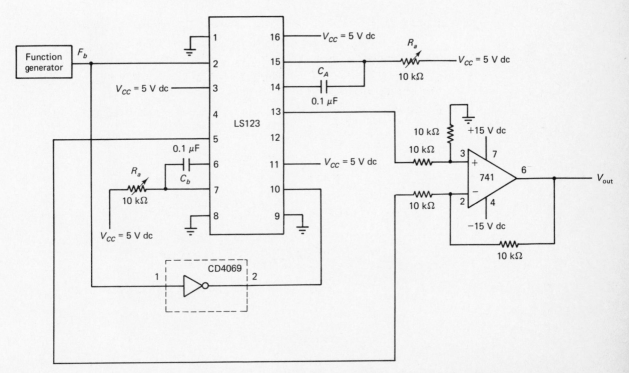

FIGURE 38-4   UPNRZ to BPRZ encoder circuit.

Procedure

1. Construct the UPNRZ to BPRZ line encoder circuit shown in Figure 38-4.
2. Use the function generator to simulate an 8000 bps UPNRZ input signal (F_b) with an alternating 1/0 sequence by setting the amplitude of the function-generator output voltage to a 5 Vp-p 4000-Hz square wave ($V_{max} = +5$ V and $V_{min} = 0$ V).
3. Adjust R_a for a 25% duty cycle rectangular wave at pin 13 of the LS123.
4. Adjust R_b for a 25% duty cycle rectangular wave at pin 5 of the LS123.
5. Sketch the input UPNRZ and output BPRZ waveforms.
6. Measure the input voltage level for a logic 1. A logic 0.
7. Measure the output voltage level for a logic 1. A logic 0.
8. Measure the average dc component of the input waveform. The output waveform.
9. Vary first R_a and then R_b and describe what effect varying them has on the shape and dc component of the output waveform.
10. Determine the ideal Nyquist frequency for the input waveform. The output waveform.

SECTION D   Summary

Write a brief summary of the concepts presented in this experiment on PCM-TDM line encoding. Include the following items:

1. The concept of line encoding.
2. The transmission characteristics of return-to-zero and nonreturn-to-zero waveforms.
3. The transmission characteristics of unipolar and bipolar waveforms.
4. The relationship between the different types of line encoding to the average dc component, bandwidth, and duty cycle of the output waveform.

EXPERIMENT 38 ANSWER SHEET

NAME: _____ CLASS: _____ DATE: _____

SECTION A

6.

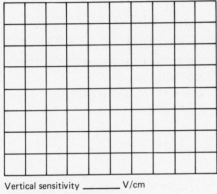 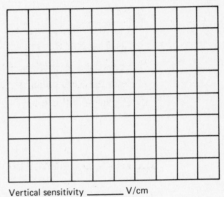

Vertical sensitivity _____ V/cm   Vertical sensitivity _____ V/cm

Time base _____ sec/cm   Time base _____ sec/cm

7. Logic 1 = _____ Logic 0 = _____   8. Logic 1 = _____ Logic 0 = _____

9. Input dc = _____ Output dc = _____ 10. F_N = _____

SECTION B

5.

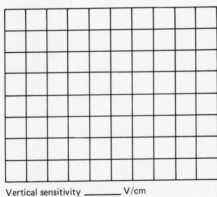 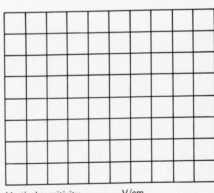

Vertical sensitivity _____ V/cm   Vertical sensitivity _____ V/cm

Time base _____ sec/cm   Time base _____ sec/cm

6. Logic 1 = _____ Logic 0 = _____   7. Logic 1 = _____ Logic 0 = _____

8. Input dc = _____ Output dc = _____

9. _____

_____

_____

10. F_N = _____

SECTION C

5.

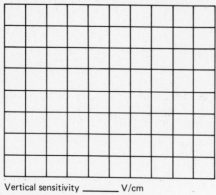 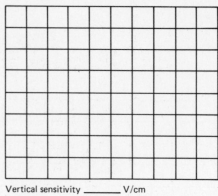

Vertical sensitivity _____ V/cm

Time base _____ sec/cm

Vertical sensitivity _____ V/cm

Time base _____ sec/cm

6. Logic 1 = _____ Logic 0 = _____   7. Logic 1 = _____ Logic 0 = _____

8. Input dc = _____ Output dc = _____

9. _____

_____

_____

10. F_N = _____

BINARY SUBSTITUTION ENCODERS

REFERENCE TEXT: Electronic Communications Systems: Fundamentals through Advanced

1. Chapter 17, T1 Carrier Systems.
2. Chapter 17, T2 Carrier Systems.
3. Chapter 17, T3 Carrier Systems.

OBJECTIVES

1. To observe the operation of a binary substitution encoder.
2. To design a binary substitution encoder.

INTRODUCTION

In this experiment the operation of a binary substitution encoder is examined. In high-speed digital carrier systems, such as the Bell System T2 and T3 PCM-TDM systems, clocking information (timing) is removed directly from the received data. Therefore, it is necessary that the data has sufficient transitions to maintain synchronization. Bipolar return-to-zero alternate mark inversion (BPRZ-AMI) encoding is used with both the T1 and T2 carrier systems. BPRZ-AMI transmissions have transitions in the data whenever 1s are transmitted. However, if the transmitted data has a long string of successive 0s, there are no transitions for the receiver to synchronize to, and clocking information is lost. For example, with the T2 carrier system, binary six zero substitution (B6ZS) encoding is used to ensure that no more than five successive 0s occur. With the T3 system, binary three-zero substitution (B3ZS) encoding is used to ensure that no more than two successive 0s occur. Binary substitution encoding involves detecting a given number of successive binary 0s and then replacing that sequence of 0s with a given code. Similarly, for other line encoding schemes, successive 1s are detected and replaced with a given code. B3ZS and B6ZS are used with BPRZ-AMI encoding schemes where each successive logic 1 transmitted has the opposite polarity from the previous one. The substituted code produces bipolar violations which are detected in the receiver. After detection, the receiver can substitute the appropriate string of consecutive 0s back into the data stream and restore the original information. In this experiment three simple (although impractical) unipolar binary substitution encoding schemes are examined. The three encoding schemes are impractical because with unipolar transmission, bipolar violations cannot occur. Therefore, it is impossible for the receiver to detect substituted patterns and restore the original information.

MATERIALS REQUIRED

Equipment

1 — protoboard
1 — dc power supply (+12 V dc)
2 — function generators (10 kHz each)
1 — standard oscilloscope (10 MHz)
1 — assortment of test leads and hookup wire

Parts list:

2 — CD4027B dual JK flip-flops
2 — CD4081B quad two-input AND gates
1 — CD4069B hex inverter

SECTION A   The Binary Four-Zero Substitution (B4ZS) Encoder

In this section the operation of a unipolar nonreturn-to-zero binary four zero substitution (B4ZS) encoder is examined. With B4ZS encoding, whenever four successive 0s are detected, they are replaced with a 1001 binary sequence. Therefore, a continuous string of 0s will generate a repetitive 100110011001 pattern. The schematic diagram for the B4ZS encoder circuit used in this section is shown in Figure 39-1a, and a sample timing diagram is shown in Figure 1b. The four JK flip-flops act as a 4-bit shift register. The parallel outputs from the shift register are monitored, and whenever four successive 0s are detected, flip-flops 1 and 4 are set and flip-flops 2 and 3 are reset.

Procedure

1. Construct the B4ZS encoder circuit shown in Figure 39-1a.
2. Use a function generator to simulate an 8000-Hz clock by setting the amplitude of the function-generator output voltage to a 10 Vp-p 8000-Hz square wave (i.e., V_{max} = 10 V and V_{min} = 0 V).
3. Simulate a continuous string of binary-0 input bits by connecting the input bits by connecting the input to the shift register to ground.
4. Sketch the clock waveform and the input and output binary waveforms observed at data in and data out, respectively.
5. Simulate a continuous string of binary-1 input bits by connecting the input to the shift register to VDD.
6. Sketch the clock waveform and the input and output binary waveforms observed at data in and data out, respectively.
7. Simulate a repetitive binary input sequence of 1111000011110000 by connecting the input to the shift register to a second function generator. Set the amplitude of the

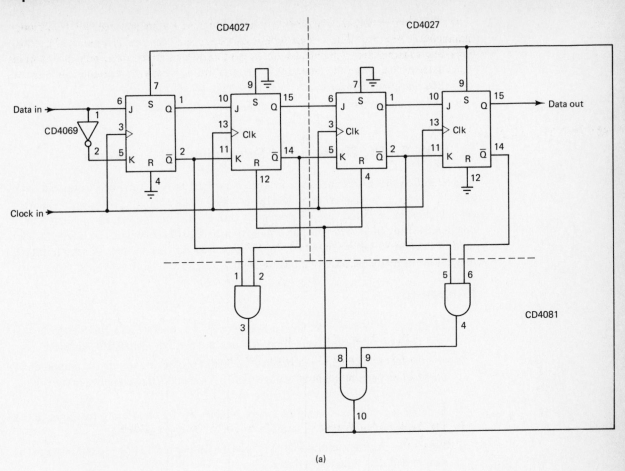

(a)

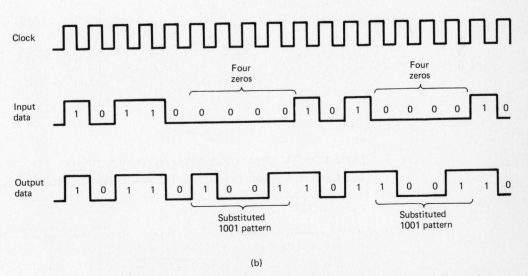

(b)

FIGURE 39-1 Binary four zero substitution (B4ZS) circuit. (a) Schematic diagram. (b) Sample timing diagram.

second function-generator output voltage to a 10 Vp-p 1000-Hz square wave (i.e., $V_{max} = 10$ V and $V_{min} = 0$ V).

8. Sketch the clock waveform and the input and output binary waveforms observed at data in and data out, respectively (it may be necessary to fine tune the second function generator to observe a stable output waveform).

9. Do not disassemble this circuit, because it is needed in Section B.

SECTION B   The Binary Four 1 Substitution (B41S) Encoder

In this section the B4ZS encoder constructed in Section A is modified to operate as a unipolar binary four 1 substitution (B41S) encoder. With encoding, whenever four successive ones are detected, they are replaced with a 1010 binary sequence. Therefore, a continuous string of 1s will generate a repetitive 10101010 sequence. The B4ZS encoder used in Section A will be modified slightly and used for this section. A sample timing diagram for a B41S encoder is shown in Figure 39-2.

Procedure

1. Modify the B4ZS encoder shown in Figure 39-1 to operate as a B41S encoder that will detect four successive 1s and generate a repetitive 10101010 sequence.

2. Use the function generator to simulate an 8000-Hz clock by setting the amplitude of the function generator output voltage to a 10 Vp-p 8000-Hz square wave (i.e., $V_{max} = 10$ V and $V_{min} = 0$ V).

3. Simulate a continuous string of binary 0 input bits by connecting the input to the shift register to ground.

4. Sketch the clock waveform and the input and output binary waveforms observed at data in and data out, respectively.

5. Simulate a continuous string of binary 1 input bits by connecting the input to the shift register to VDD.

6. Sketch the clock waveform and the input and output binary waveforms observed at data in and data out, respectively.

7. Simulate a repetitive binary input sequence of 1111000011110000 by connecting the input to the shift register to a second function generator. Set the amplitude of the second function generator to a 10 Vp-p 1000-Hz square wave (i.e., $V_{max} = 10$ V and $V_{min} = 0$ V).

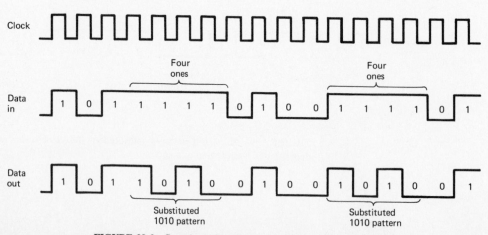

FIGURE 39-2   Sample timing diagram - binary four one substitution.

8. Sketch the clock waveform and the input and output binary waveforms observed at data in and data out, respectively (it may be necessary to fine tune the second function generator to observe a stable output waveform).

SECTION C   Summary

Write a brief summary of the concepts presented in this experiment on binary substitution encoders. Include the following concepts:

1. Clock synchronization.
2. The importance of transitions in the transmitted data waveform.
3. The purpose of binary substitution.
4. The operation of a binary four-0 substitution encoder and a binary four-1 substitution encoder.

EXPERIMENT 39 ANSWER SHEET

NAME: _____ CLASS: _____ DATE: _____

SECTION A

4.

Vertical sensitivity _____ V/cm

Time base _____ sec/cm

Vertical sensitivity _____ V/cm

Time base _____ sec/cm

Vertical sensitivity _____ V/cm

Time base _____ sec/cm

6.

Vertical sensitivity _____ V/cm

Time base _____ sec/cm

Vertical sensitivity _____ V/cm

Time base _____ sec/cm

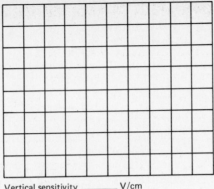

Vertical sensitivity _____ V/cm

Time base _____ sec/cm

8.

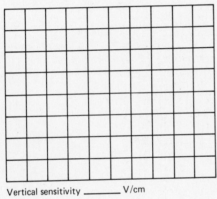

Vertical sensitivity _____ V/cm

Time base _____ sec/cm

Vertical sensitivity _____ V/cm

Time base _____ sec/cm

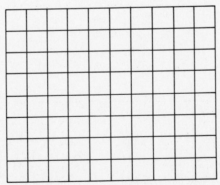

Vertical sensitivity _____ V/cm

Time base _____ sec/cm

SECTION B

4.

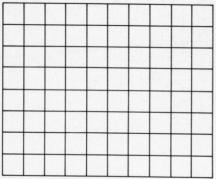

Vertical sensitivity _____ V/cm

Time base _____ sec/cm

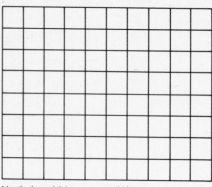

Vertical sensitivity _____ V/cm

Time base _____ sec/cm

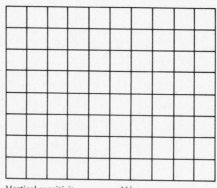

Vertical sensitivity _____ V/cm

Time base _____ sec/cm

7.

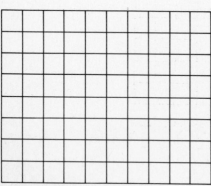

Vertical sensitivity _____ V/cm

Time base _____ sec/cm

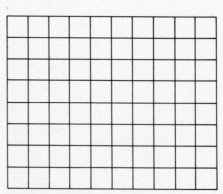

Vertical sensitivity _____ V/cm

Time base _____ sec/cm

Vertical sensitivity _____ V/cm

Time base _____ sec/cm

9.

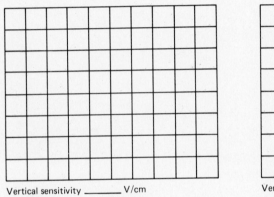

Vertical sensitivity _____ V/cm

Time base _____ sec/cm

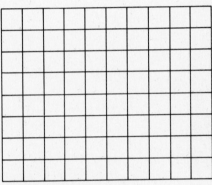

Vertical sensitivity _____ V/cm

Time base _____ sec/cm

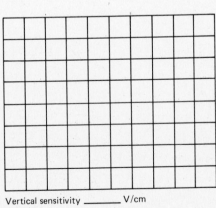

Vertical sensitivity _____ V/cm

Time base _____ sec/cm

BIPOLAR RETURN-TO-ZERO ALTERNATE-MARK INVERSION ENCODER/DECODER DESIGN

REFERENCE TEXT: Electronic Communications Systems: Fundamentals through Advanced

1. Chapter 17, Line Encoding.

OBJECTIVES

1. To observe the operation of a bipolar return-to-zero alternate-mark inversion (BPRZ-AMI) line encoder/decoder.
2. To design a BPRZ-AMI line encoder/decoder to a given set of specifications.
3. To observe the transmission characteristics of a BPRZ-AMI encoded signal.

INTRODUCTION

Line encoding involves converting standard logic levels (i.e., TTL, CMOS, etc.) to a form more suitable for telephone line transmission. Essentially, there are four primary factors that must be considered when selecting a line encoding format.

1. Timing (clock) recovery. With PCM/TDM carrier systems, clocking information is generally recovered directly from the received data. Therefore, there must be a sufficient number of transitions in the data signal to allow for continuous and accurate clock recovery.

2. Transmission bandwidth. The minimum bandwidth required to propagate a digital line encoded signal through a transmission medium is determined by the highest fundamental frequency present in the waveform. The highest fundamental frequency is determined from the worst-case (fastest-transition) binary bit sequence. The highest fundamental frequency is the minimum bandwidth and is called the minimum or ideal Nyquist frequency.

3. Ease of detection and decoding. The dc component associated with a digital line encoding scheme and the transmit voltage levels are dependent on the particular line-

encoding scheme used. Unipolar transmission transmits 0 volts for a logic 0 and either a positive or a negative voltage for a logic 1. Bipolar transmission transmits both positive and negative voltages for logic 1s and zero or negative voltages for logic 0s. The average dc component for a digital transmission is dependent on whether unipolar or bipolar encoding is used.

4. Error detection. With the rapidly increasing transmission bit rates, error detection has become an important consideration when selecting a line encoding scheme.

In this experiment a bipolar return-to-zero alternate-mark inversion (BPRZ-AMI) encoder/decoder circuit is designed, constructed, and tested to meet a prescribed set of specifications; then the transmission characteristics of BPRZ-AMI are examined. The block diagram for the BPRZ-AMI encoder/decoder circuit used in this experiment is shown in Figure 40-1.

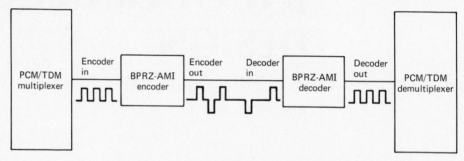

FIGURE 40-1   BPRZ-AMI encoder/decoder block diagram.

MATERIALS REQUIRED

Equipment:

1 — protoboard

1 — dual dc power supply (-15 V dc to $+15$ V dc)

2 — function generators (100 kHz each)

1 — standard oscilloscope (10 MHz)

1 — assortment of test leads and hookup wire

Parts list:

1 — assortment of resistors, capacitors, and integrated circuits depending on the individual circuit implementation.

SECTION A   Line-Encoding Transmission Theory

In this section the transmission theory for BPRZ-AMI line encoding is examined. With unipolar transmission, a single represents a logic 1 (i.e., either $+V$ or $-V$, depending on whether positive or negative logic is used), and 0 volts represents a logic 0. For example, with standard TTL a logic $1 = +5$ V and a logic $0 =$ ground. With bipolar transmission, a logic 1 can be represented by both positive and negative voltages (i.e., logic $1 = +V$ or $-V$), and zero volts represents a logic 0; or logic 1s can be represented with a positive voltage ($+V$) and logic 0s represented with a negative voltage ($-V$). With nonreturn-to-zero transmission, the level of the binary pulse is maintained for the entire bit time (i.e., a 100% duty cycle); and with return-to-zero transmission the active time of the binary pulse is less than 100% of the bit time and the pulse returns to V (ground) for the remainder of the pulse time (generally, a 50% duty cycle is used).

Unipolar nonreturn-to-zero (UPNRZ) is the standard binary transmission scheme used internally by computers and conventional digital equipment. With conventional UPNRZ encoding, a logic 1 = +V and a logic 0 = 0 V; and a 100% duty cycle pulse is transmitted. Unipolar return-to-zero (UPRZ) is identical to UPNRZ except that a 50% duty cycle is used. Bipolar nonreturn-to-zero (BPNRZ) encoding involves transmitting a positive voltage for a logic 1 and a negative voltage for a logic 0, and a 100% duty cycle pulse is used. Bipolar return-to-zero (BPRZ) encoding is identical to BPNRZ encoding except that a 50%-duty-cycle pulse is used. With bipolar return-to-zero alternate mark inversion (BPRZ-AMI) encoding, there are two nonzero voltages (+V and −V), but both polarities represent a logic 1; 0 V represents a logic 0. With AMI transmissions, each successive logic 1 is inverted in polarity from the previous logic 1. BPRZ-AMI encoding has a built in error-detection mechanism. When a transmission error occurs—for example, when a transmitted logic 1 is changed to logic 0 or vice versa—the receiver will detect a bipolar violation (i.e., the reception of two successive logic 1s with the same polarity). Once transmission errors are detected, higher-level procedures can be implemented to restore the integrity of the system. Sample timing diagrams for UPNRZ, UPRZ, BPNRZ, BPRZ, and BPRZ-AMI transmissions are shown in Figure 40-2.

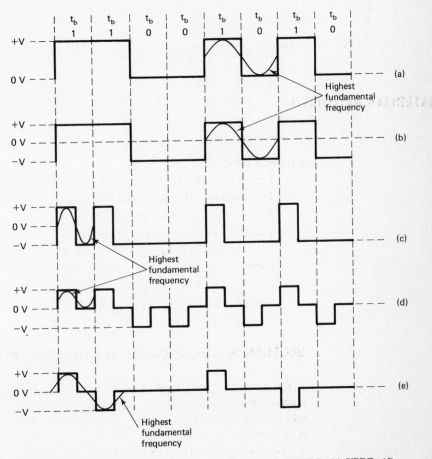

FIGURE 40-2 Line-encoding formats: **(a)** UPNRZ; **(b)** BPNRZ; **(c)** UPRZ; **(d)** BPRZ; **(e)** BPRZ-AMI.

SECTION B   Design Specifications

In this section the design specifications for the BPRZ-AMI encoder/decoder circuit that will be designed in this experiment are given. The system must be designed to operate in the simplex (one-way only) mode. Therefore, only one encoder and one decoder will be constructed and tested. The binary signal source and external clock circuit can both be simulated with function generators. The line encoder-decoder specifications are given in Table 40-1 (refer to Figure 40-1 for encoder input and output signals).

TABLE 40-1   BPRZ-AMI Design Specifications

	Encoder input and decoder output	Encoder output and decoder input
Encoding	Standard TTL	BPRZ-AMI
Logic 1 level	<5 V	>+10 V and <−10 V
Logic 0 level	<0.2 V	<0.2 V
Duty cycle	100%	50% ± 10% of bit interval
Rise and fall times	-------------------------------------- <5% of bit interval --------------------------------------	
Bite rate	-- 10 kbps --	
Transmission meduim	------------------------------------- 18 to 24 guage wire pair -------------------------------------	
Transmission distance	-- >10 feet --	

SECTION C   BPRZ-AMI Encoder Design

In this section a BPRZ-AMI encoder circuit will be designed, constructed, and tested to meet the specifications given in Table 40-1. The only restriction on what components you can use is that a single chip especially designed for BPRZ-AMI encoding/decoding cannot be used.

Procedure

1. Design an encoder circuit that will convert standard TTL levels to the BPRZ-AMI specifications given in Table 40-1.
2. Construct the circuit designed in step 1.
3. Test the BPRZ-AMI encoder circuit constructed in step 2 to determine if the encoder is operating within the design specifications given in Table 40-1.
4. Troubleshoot the BPRZ-AMI encoder circuit tested in step 3, and make any design changes, adaptions, or additions that are necessary for the encoder to operate within the given specifications.

SECTION D   BPRZ-AMI Decoder Design

In this section a BPRZ-AMI decoder circuit will be designed, constructed, and tested to meet the specifications given in Table 40-1. The only restrictions on what components you can use is that a single chip especially designed for BPRZ-AMI encoding/decoding cannot be used.

Procedure

1. Design a decoder circuit that will convert the BPRZ-AMI signal specified in Table 40-1 to standard TTL levels.

2. Construct the BPRZ-AMI decoder circuit designed in step 1.
3. Test the decoder circuit constructed in step 2 to determine if the decoder is operating within the design specifications given in Table 40-1.
4. Troubleshoot the BPRZ-AMI decoder circuit tested in step 3, and make any design changes, adaptions, or additions that are necessary for the decoder to operate within the given specifications.

SECTION E   Circuit Implementation

In this section the BPRZ-AMI encoder and decoder circuits designed, constructed, and tested in Sections C and D are put together and tested as a system. Make any adjustments and changes necessary for the system to operate within the given specifications.

SECTION F   Summary

Write a complete laboratory report for the BPRZ-AMI encoder/decoder system designed in Sections C, D, and E. Include the following items:

1. Complete schematic diagrams of the encoder and the decoder.
2. An introduction, including a statement of the problem.
3. All pertinent data.
4. A list of problems encountered and how they were rectified.
5. A conclusion.